The biogeography of the British Isles

RV

The current interest in ecology and the environment has highlighted the value of biogeography, which approaches both the natural and the man-made environment from the perspective of spatial distribution. This is the first book devoted to the biogeography of the British Isles and surrounding shelf seas. Bringing together a wealth of diverse information, it is thoroughly referenced and well illustrated, and will be invaluable to students of geography, environmental science, ecology, botany, and zoology.

Peter Vincent traces the development of British biogeography over the last two centuries, examining key topics such as ecosystems, habitats, and niches in the context of plant and animal distribution. He gives a detailed account of the development of biogeographical mapping and recording systems, and describes modern-day distributions, both in the countryside and in urban areas against the backcloth of disruptive human activities.

The Author
Peter Vincent is Lecturer in Geography at Lancaster University.

The biogeography of the British Isles
An Introduction

Peter Vincent

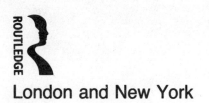

ROUTLEDGE

London and New York

First published 1990
by Routledge
11 New Fetter Lane, London EC4P 4EE

Simultaneously published in the USA and Canada
by Routledge
a division of Routledge, Chapman and Hall, Inc.
29 West 35th Street, New York, NY 10001

Typeset by J&L Composition Ltd, Filey, North Yorkshire
Printed and bound in Great Britain by
Mackays of Chatham PLC, Chatham, Kent

British Library Cataloguing in Publication Data

Vincent, Peter, 1944–
  The biogeography of the British Isles
  1. Biogeography
  I. Title
  574.9

  ISBN 0-415-03470-1
       0-415-03471-X (pb)

*Library of Congress Cataloging in Publication Data
has been applied for*

Contents

Figures

Tables

Preface

There are few subjects more fascinating than biogeography. In many respects it can be thought of as at the 'heart of geography'. This may seem a rather presumptuous statement, although others before have made the same claim. Whether we like it or not, most of our lives are intricately linked in one way or another with the world of plants and animals. We must sometimes look carefully to see that this is so, however, particularly if we live in an urban area with acres of concrete.

It seems to me that one of the fundamental problems with modern geography, and, indeed, much of biogeography, is its emphasis on rather abstract processes, with very little attention being paid to accounts of spatial variation and location. Perhaps some of this mistaken emphasis is due to the quite false idea that the study of distributions (and regions, for that matter) is rather trivial and not very scientific. Indeed the clamour for so-called rigour and the adoption of process and systems studies by geographers in the 1960s has led, in my view, to a partial loss of identity for the discipline as a whole. This trend has particularly affected bio-geography, such that many geographical texts hardly distinguish between ecology and biogeography.

There is a very real case for the study of biogeography being centred on distribution studies. One of the most compelling arguments is the very real need for spatial bench-marks, without which it is almost impossible to evaluate the impacts of natural and man-made environmental changes on fauna and flora. How can we assess, for example, the possible bio-geographical impacts of acid rain, global warming, and urbanization without the fundamental distributional data?

Writing a book about the biogeography of the British Isles and sur-rounding seas is fraught with difficulties and contradictions. On the one hand, some of the biogeography is well-known, although frequently buried in obscure journals. This is true, for example, of plants and most classes of the larger animals. On the other hand, it is probably true to say that the biogeography of the majority of invertebrates is poorly known, and for many such animals, not known at all. The distribution of many marine organisms is, similarly, only known in the broadest outline.

In the last 30 years or so the systematic recording of biogeographical data in pre-defined map grid-squares is an important scientific advance, although for some parts of the British Isles, such as Eire, the coverage is not as good as it might be. Grid data are both easily stored and manipulated by computer, and are an invaluable but under-used source of biogeographical information. Where possible, maps in the text have been drawn using such data.

Doubtless, much more could have been said about the biogeography of the shelf seas surrounding the British Isles. Quite naturally, the seas are less well known than the land, and are intrinsically more difficult to study. Nevertheless, I hope I have included enough information to deal with the major factors affecting marine distributions.

The ten chapters in the book fall naturally into four sections. In Chapters one to three I have tried to set the scene in terms of how biogeography and its methods have evolved in the British context, and also in terms of the essential characteristics of the British environment. The next three chapters constitute a core of basic biogeographical information, with particular emphasis on those factors influencing plant and animal distributions. In Chapters seven and eight, biogeographical ranges and their evolution during the Tertiary and Quaternary periods are described in depth. Finally, two chapters are devoted to the biogeography of rural and urban man-made landscapes.

Acknowledgements

I am indebted to a great many people who have helped me in the writing and preparation of this book. I am particularly pleased to thank Dr Robert Jones and Dr David Bartley, both of whom took on the onerous task of reading and commenting on the complete text. I should also like to acknowledge the painstaking work of the cartographers in the Geography Department at Lancaster University: a special vote of thanks goes to Claire Jarvis and Peter Mingins (alas, now departed for the world of big pay packets outside the giddy heights of academia). Although they did not know it at the time, at least some of my ideas for this book were tried out on Jessie and Lois on those enjoyable days botanizing on the Lancashire and Cumbrian fells. My final thanks are to Louise, without whose total support nothing at all would have been achieved.

I would like to thank the following for their kind permission to reproduce material from previously published work: Academic Press Inc. (London) Ltd for Figure 7.7; Edward Arnold for Figure 3.11; Biological Records Centre for Figures 3.1, 3.2, 3.9, 3.14, 6.2, 6.3, 6.7, 9.4, and 10.3; Blackwell Scientific Publications Ltd for Figures 4.2 and 4.12; British Trust for Ornithology for Figures 5.7a and 5.7b; Cambridge University Press for Figures 4.4, 4.15, 5.3, Tables 4.2 and 7.3; Chapman and Hall for Figures 4.7 and 5.11; Classey Ltd for Table 9.3; Conchological Society of the British Isles for Figure 6.9; Countryside Commission for Figure 9.5; J. G. Dony for Figure 3.7; Field Studies Council for Figure 2.15; Geological Society of America for Figure 9.1; Ministry of Agriculture, Fisheries, and Food for Figures 2.5, 2.6, and 2.7; Nature Conservancy Council for Figure 9.7; Pergamon Press Ltd for Figure 8.11; Oxford University Press for Figures 2.3, 7.9, and Table 1.2; R. F. Tait for Figures 2.2 and 2.4; Liverpool University Press for Figure 10.4; Longman Group Ltd for Figures 2.8, 2.9, 2.10, 2.12, 2.13, 2.16, 2.17, and 10.1; Nature Conservancy Council for Table 9.2; Societas Biologica Fennica Vanamo for Figures 7.1, 7.2, 7.3, and 7.4; Editor, Scottish Geographical Magazine, for Table 2.3; Springer-Verlag for Figure 2.18.

Chapter one

The history and scope of British biogeography

Introduction

Biogeography is a strange discipline. In general, there are no institutes of biogeography; there are no departments of it. There are no professional biogeographers – no professors of it, no curators of it. It seems to have few traditions.

This quotation is from a paper on the history of biogeography by the American palaeontologist Gareth Nelson (1978). Although one might quibble with details (there are now, for example, biogeographers holding Chairs in several British geography departments), the picture painted by Nelson is, to all intents and purposes, correct. Despite this gloomy scenario, however, biogeography is a flourishing discipline – studied not only by geographers but also by biologists and geologists. This diversity of interest is part strength but also part confusion, especially for those coming to the subject for the first time.

The view taken here is that *biogeography is the study and interpretation of plant and animal distributions in geographic space*. Such a simple definition raises all sorts of difficult questions and issues. What scale of geographical space is relevant – the world, Belfast, Hyde Park, a garden, a leaf? Is it relevant for the animal, the plant, or the biogeographer? Should biogeography be concerned with the distribution of individual species or assemblages of species? Is there a legitimate distinction between the study of plant dispersal aided by water currents and virus dispersal aided by the movements of man acting as host? The answers to these and similar questions have yet to be resolved, if indeed they can be resolved at all. We might also bear in mind that geographic space is three-dimensional but only rarely have biogeographers examined distributions other than on two-dimensional maps. Sometimes such maps reveal little about the total activity space of the organism; this is particularly true of marine animals and birds.

Modern biogeography is best viewed as a broad interdisciplinary subject with many sub-fields, such as: zoogeography (the study of animal distributions); phytogeography (the study of plant distributions); and historical biogeography (the study of distributions over time). Within each sub-field there are sometimes widely varying methodological approaches. Yet despite this apparent diversity, biogeography does seem to have an essential cohesiveness which is provided by the simple fact that living organisms are not uniformly distributed over the face of the earth. To reveal the processes that give rise to this variation remains a fundamental goal.

Before describing contemporary paradigms in British biogeography we shall first look at the way it has emerged as an academic discipline over the last couple of centuries. Biogeography has produced some of the big questions of science and has deservedly attracted a number of distinguished scientists, such as Lyell, Darwin, and Wallace. The historical perspective is important because a number of simple misconceptions about biogeography abound. To judge by the majority of recent texts written on biogeography we might be led to conclude that biogeography was, more or less, just the study of ecosystems. But this view ignores the very important role of the spatial description and analysis. It also tends to ignore the use of the map as a biogeographical tool in its own right. Instead we are fed on a diet of systems diagrams, feedback loops, and so on; all, of course, very useful but perhaps more central to ecology than biogeography. It is also important to understand that biogeography has always been very much the domain of biology rather than geography in terms of published books, research papers and major scientific advances. Indeed, when the first university geography departments came into being in the late nineteenth century, biogeography was already a well established, although rather loosely organized, science.

A number of excellent studies on the history of biogeography have been written and are well worth delving into. Probably the most useful are *The Secular Arc* (Browne 1983) and the concluding chapter in *On Geography* (Stoddart 1986). There are also comprehensive essays in *Themes in Biogeography* which is a major statement on biogeography as seen through the eyes of a group of British geographers (Taylor 1984). Cruickshank (1984) provides a recent overview of the history of biogeography in Ireland.

A particularly interesting book is *The Background of Ecology* (McIntosh 1985) in which the biogeographical origins of ecology are briefly described. A snippet from McIntosh's fascinating book illustrates the point that learning about the history of a subject can be both entertaining as well as educative. In discussing a book by the famous plant geographer, H. C. Watson, McIntosh suggests a degree of irritation among nineteenth-century feminists when they might have read in a footnote:

It may not be amiss to observe for the benefit of lady readers, or others who are not familiar with Greek and Latin names, that the one adopted for the present work, is to be pronounced in three syllables, thus, 'cy-be-le'.

(Watson 1847: 69)

No room in the Ark?

Prior to the mid-eighteenth century, British naturalists had little reason to think that different geographical areas would be occupied by different species. At that time it was commonly supposed that the world had been overcome by the waters of the great biblical Deluge. After the Deluge subsided, the living organisms saved in Noah's Ark dispersed over an unoccupied globe from the Ark's resting place on Mount Ararat in eastern Turkey. But with more and more biological specimens being brought back from discoveries both in the New and Old Worlds and the realization that there were major regions of the earth with distinctive plant and animal species adapted to their environment, the literal biblical explanation of floral and faunal distributions became untenable (Kinch 1980).

As knowledge of the number of plant and animal species grew, naturalists also began to puzzle over the size of the Ark. For example, John Ray, Britain's premier seventeenth-century naturalist was familiar with some 500 species of birds and approximately 10,000 invertebrates. The discoveries also brought other problems: how, for example, did reindeer and polar bears manage to disperse from the Ark across arid deserts to the Polar region? The absurdity of a global biogeography based on Noah's Ark was recognized, and it provided the necessary stimulus for the production of more scientific alternative explanations of biogeographical distributions and dispersal.

A grand design or natural laws?

In the first half of the nineteenth century the study of biogeography had attracted a good deal of interest among the great naturalists and geologists of the day. By the early 1800s many had begun to question the efficacy of the Deluge and this led to the acceptance of the concept of several regions of creation (*sensu*, the Bible). But the acceptance of this concept led to one of the major biogeographical issues of the nineteenth century – the problem of disjunct distributions. These are distributions in which populations of a species are widely separated in geographical space. In order to explain disjunct distributions, several differing theoretical approaches were taken. One approach simply stated that the presence of a species in two widely separated locations implies two separate creative acts. An alternative explanation supposed that a species had been created only in one place and the disjunction was due to the migration of some part of the population from the place of origin.

3

By the early nineteenth century, then, there were two major bio-geographical issues at stake: one was the enumeration and delineation of the regions of creation; the other was the question of disjunct distributions. During the 1840s, British biogeographical theorists had split into two camps. On the one side there were the catastrophists, such as the ornithologist P. L. Sclater, who saw a world neatly planned and organized by the hand of God at creation. On the other, the opposition saw geographical distributions as being dynamic, changing through time and controlled by natural laws. They were expounding the ideas of uniformi-tarianism, the view expounded by James Hutton (1726–97) that 'the present is the key to the past' (Kinch 1980).

In 1858, Sclater published a theoretical paper on the geographical delineation of life based upon birds. On the basis of the known distribution of perching birds (passerine species) he divided the world into two major regions – Neogea (the New World) and Palaeogea (the Old World) and continued the subdivision until he arrived at six major regions which were later to be adopted by A. R. Wallace (Figure 1.1). For Sclater, distributions did not change with time and he simply dismissed disjunct distributions, arguing that each disjunct population was a separate super-natural creative act because he did not believe that any geological forces could widely separate two populations of a species. Sclater and his followers saw no reason for migrations because God had designed species to be perfectly adapted to their environments. For Sclater, the identifica-tion of geographical regions were the revelation of God's idealist design which would hold equally for all taxonomic groups, even man.

The famous geologist Charles Lyell (1797–1875) put forward his opposi-tion to the supernatural explanation of biogeographical distributions in the second volume of his *Principles of Geology* (1830–33). Lyell was out to 'sink the diluvialists, and in short, all theological sophists', and replaced the idea of a biogeographical clean slate, after a major catastrophe such as the Deluge, with the notion of a steady-state earth governed by one uninterrupted succession of geological events controlled by known laws. For Lyell, the so-called 'regions of creation' were simply the result of dispersal over time being checked by natural barriers. Moreover, geo-graphical distributions were thought by Lyell to be dynamic, expanding and contracting their boundaries as geological agents affected topography and climate. Lyell's writings had a profound influence on his contem-poraries such as Forbes, Hooker, Wallace, and Darwin. Encouraged by Lyell's attempt to integrate geological movements with distribution, Forbes (1815–54) postulated the spread of plants and animals over large areas which are no longer connected together – the idea of the land-bridge was born.

Edward Forbes was born on the Isle of Man and is now particularly remembered for his memoir 'On the connexion between the distribution of

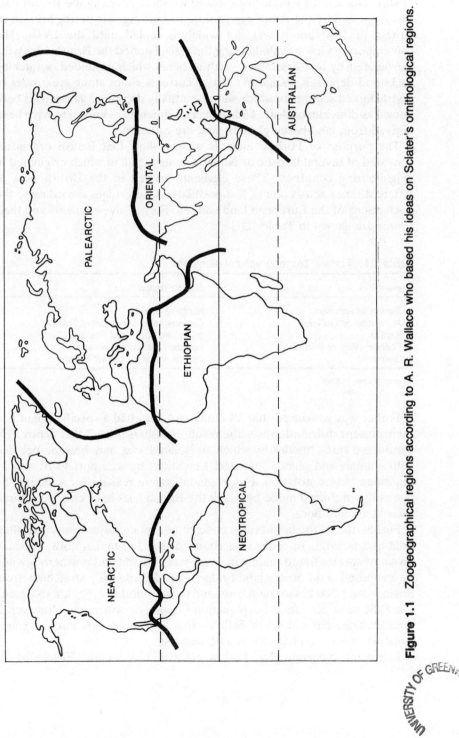

Figure 1.1 Zoogeographical regions according to A. R. Wallace who based his ideas on Sclater's ornithological regions.

existing fauna and flora of the British Isles, and the changes which have affected their area, especially during the epoch of the northern drift' (1846). The idea of a major land-based ice sheet covering the British Isles was suggested by the American geologist Louis Agassiz during his visit to Britain in 1840 but it was not widely accepted until the 1860s. The contemporary view was that during the glacial period the British Isles were surrounded by icy seas replete with icebergs which grounded, scratching rocks and depositing boulders. Strong currents swept along great tides of poorly mixed sediments which were eventually deposited as jumbled beds known as diluvium or drift. Today, such sediments are known to have been derived from land-based glaciers, and are called tills.

The purpose of Forbes' memoir was to show that British organisms consisted of several floristic or faunal 'elements', all of which originated in neighbouring countries. These elements arrived in the British Isles at different times across one of five postulated land-bridges according to the fluctuations of the European land and sea. Forbes' five elements and their source are shown in Table 1.1.

Table 1.1 Forbes' biogeographical elements

Element	Source region
1. Iberian or Asturian	northern Spain
2. Armorican or Gallician	Channel Isles and western France
3. Kentish	north and north-east France
4. Scandinavian or Boreal	northern and sub-arctic area
5. Germanic	central and west central Europe

Source: Forbes (1846)

Forbes was convinced that all fauna and flora had a specific origin and their present distribution was the result of dispersal from that centre. He considered three 'modes by which an isolated area may become peopled with animals and plants': by special creation; by transport to it; and by migration before isolation. He presented cogent reasons for selecting the last as the preferred mode by which the British Isles have chiefly acquired their fauna and flora.

Forbes' use of the land-bridge concept to explain disjunctions was often bold and fanciful. In order to account for the Lusitanian flora – a small group of species found mainly in south-west Ireland and northern Spain – he proposed a Miocene land-bridge, called 'Atlantis', stretching from Spain some 1,500 km to the Azores and thence another 1,500 km to Ireland – not the most economical explanation I think you will agree. However, it does illustrate one of Forbes' beliefs – that disjuctions were due to a single isolating event rather than repeated dispersal.

In marine biogeography, Forbes' treatise *The Natural History of the*

European Seas (1859) remains a classic description of major marine biogeographical regions which he defined as the Arctic, Boreal, Celtic, Lusitanian, and Mediterranean provinces. Forbes was one of the founders of marine 'dredging' studies of benthic organisms, and linked his biological observations of sea-bottom invertebrates to his ideas of geological and environmental change.

Not all biogeographical study at this time was concerned with processes. There were naturalists with a taxonomic rather than a geological background who were beginning to investigate the existence and arrangement of groups of living organisms. In effect, geologists were concerned with distributions over time, while the natural historians attempted to describe geographical distribution as it occurred in space, particularly in relation to climate. This approach is well illustrated in W. J. Hooker's *Geography Considered in Relation to the Distribution of Plants* (1834).

A leading figure among the so-called topographical biogeographers was the botanist Hewett C. Watson (Egerton 1979). Watson (1804–81) was undoubtedly influenced by the French botanist Augustin de Candolle and the great German geographer Alexander von Humboldt. It was de Candolle who coined the term 'botanical arithmetic' which was a simple numerical method whereby absolute figures were transformed into ratios which could be ranked in a table so as to reveal geographical patterns. For example, de Candolle calculated that France possessed an average of 7.2 species per genus of flowering plants, whereas the British Isles had 2.3 and the Canary archipelago 1.5.

In 1832 Watson published his first book, entitled *Outlines of the Geographical Distribution of British Plants; belonging to the division of Vasculares or Cotyledones.* The book is divided into two parts: the first being a general discussion, the second providing brief descriptions of all species of British plants, giving details not only of their habitat requirements and topographical range but also their world-wide distribution. Watson classified British vegetation into three regions each of which is subdivided into zones:

   I. Wood region   1 agricultural zone
                        2 upland zone
  II. Barren region 3 moorland zone
                        4 sub-alpine zone
III. Mossy region 5 alpine zone
                        6 snowy zone

Each of the six zones was characterized by an indicator species which was common throughout that zone but absent from the zones above and below:

1 agricultural zone ends where the cultivation of wheat ceases;
2 upland zone ends where hazel (*Corylus avellana*) ceases;

7

3 moorland zone ends where the stiff sedge (*Carex bigelowii*) begins;
4 sub-alpine zone ends where ling (*Calluna vulgaris*) ceases;
5 alpine zone ends where crowberry (*Empetrum nigrum*) ceases;
6 snowy zone ends where the land terminates.

Watson's strength lay in his meticulous collection and organization of data. This organizational forte was to show through in his four-volume *Cybele Britannica; or British Plants and their Geographical Relationships* (1847–59) in which he divided Britain into 112 recording areas called vice-counties (see Chapter three). This one invention was to have a profound influence on biogeographical studies up until the present day. The spatial distributions of plants, and animals for that matter, could now be examined in detail and explanatory hypotheses sought. Ironically, botanists were rather slow to use Watson's mapping scheme, whereas zoologists saw its potential more readily. Taylor's classic *Monograph of the Land and Freshwater Mollusca of the British Isles* published in Leeds in 1907 was among the first studies to utilize Watson's divisions and has many beautifully hand-painted distribution maps.

Darwin and Wallace

Alfred Russel Wallace (1823–1913) and Charles Darwin (1809–82) dominated biogeographical thought in the second half of the nineteenth century. Both were great travellers: Darwin to South America and the Galapagos Islands, Wallace to the Malay Archipelago. There is no doubt that this first-hand knowledge of the huge variety of plant and animal species in the tropics led both of these great men to propose a theory of evolution in papers presented to the Linnean Society of London in 1858. Until this time, biogeographers were still arguing about centres of creation (Darlington 1959). For example, Sclater, in a paper to the Linnean Society in 1857, said that:

> if he could show the existence of separate centres of creation in different parts of the world, he could avoid the awkward necessity of supposing the introduction of the red man into America by the Bering Straits, and of colonizing Polynesia by stray pairs of Malays floating over the water like cocoa-nuts, and all similar hypotheses.

Darwin's theory of evolution stopped this sort of nonsense. In his *On The Origin of Species by Means of Natural Selection* published in 1859 and in Wallace's *The Geographical Distribution of Animals* (1876) and *Island Life* (1880) a detailed case was presented for the notion that not only did plants and animals disperse over a more or less stable earth, but in so doing they themselves could alter in structure and evolve. Darwin's theory of evolution was based on a number of simple observations and deductions.

The line of reasoning goes something like this. Most organisms have the ability to produce a very large number of offspring; a single orchid, for example, might produce 1.7×10^6 seeds in a season. All organisms, even members of the same species, will vary. The offspring most likely to survive are those which have a combination of features which enable them to cope best with their environment. This latter concept is called survival of the fittest by natural selection. Lastly, the features favoured by natural selection are inherited. It is important here to note that both Darwin and Wallace discovered evolution through their biogeographical studies (Ghiselin 1969), and particularly through their studies of the dispersal capabilities of organisms (Fichman 1977). The fact that most organisms produce a great many offspring was an important one to Darwin, who thought that, given enough time, even improbable dispersal becomes probable.

According to Wilma George (1964), the subject of zoogeography in its modern form may be said to have been 'invented' by Wallace who described his aim as solving the problems of why animals live where they do. Wallace thought that all distributions could be explained by causes of four kinds: dispersal, evolution, geographical change, and climatic change. In essence, the Darwin–Wallace tradition saw the Palaearctic and Oriental regions as sources from which all other regions were supplied with higher forms of life by dispersal over fixed continents.

No account of Darwin and Wallace is complete without at least some brief mention of their contemporary, Sir Joseph Dalton Hooker, who was perhaps the world's most ambitious plant collector (Turrill 1959). His travels to many remote parts of the globe allowed him to develop the important concept of long-distance dispersal to account for those plants on remote islands with easily dispersed seeds and fruits. Hooker's plant collections were a great source of phytogeographical information and both Darwin and Wallace publically acknowledged their debt to this great scientist.

The rise of an ecological tradition

By the beginning of the twentieth century, speculative accounts of the origin and migration of species were being replaced by rigorous experimental studies of the adaptation of plants to their environments, and the physiological orientation of new texts made a compelling framework for explaining distributions (Hagen 1986). In 1904, stimulated particularly by the writings of the Danish botanist Eugenius Warming, a group of ecologists formed The Central Committee for the Survey and Study of the British Vegetation (later shortened to the British Vegetation Committee) with a view to survey all the plant communities of the British Isles. This research continued until 1913, when the Committee was dissolved and regrouped into the British Ecological Society, which in the same year launched its *Journal of Ecology*. Duff and Lowe (1981) describe in detail how this period also saw the beginnings of the conservation movement in Britain, with the formation of organizations such as the Royal Society for the Protection of Birds (founded in 1889), the National Trust (1885), and the Society for the Promotion of Nature Reserves (founded in 1912 and given its Royal Charter in 1981).

Two points are worth emphasizing. For the first time, a group of scientists had emerged who were not so much interested in individual distributions but the distributions of assemblages of plants (vegetation). At this point we ought to draw a clear distinction between the words flora and vegetation. By flora we simply mean a list of the species present in an area. If, however, we have some quantitative representation of the individual species in the flora, we can describe its vegetation. This emphasis on vegetation rather than on individual species can be thought of as the beginnings of the ecological tradition in British biogeography. A second point concerns the term ecology (Greek *oikos* – home or dwelling, which was coined by American essayist and nature-lover, H. D. Thoreau in 1858, but introduced into the scientific literature by the German biologist Haeckel in 1866. Ecology to Haeckel was the study of the interactions of all organisms living together in one locality and their adaptations to their environment. Geographers in particular were to take to ecological bio-geography because it gave them a way of integrating biological phenomena, environment, and man. It was left to others to maintain a spatial tradition which relied very much on the biogeographer also being a good taxonomist (Stoddart 1977).

In 1912 the German geophysicist Alfred Wegener (1880–1930) first described a theory of continental drift and a number of plant geographers found it useful as an explanation of disjunct distributions. Wegener's writings were translated into English but were only slowly accepted by British scientists. One British biogeographer who was quick to see the importance of the theory of continental drift was Ronald Good. In 1930 he published an early paper on the disjunct distribution of shrubby plants in the genus *Coriaria*.

The then controversial theory of continental drift was matched in the 1920s by the equally controversial 'age and area' theory of J. C. Willis (1868–1958), a prominent plant taxonomist. In his book of the same title, Willis (1922) argued that the geological age of a species was directly correlated with its area of distribution. The oldest species in a genus would tend to be most widely distributed and, by implication, endemic species were nearly always young. Willis' theory became quite notorious and generated much heated debate (Gleason 1924). It also made plant geogra-phers reconsider their views on endemism, dispersal, and patterns of speciation; for that reason Willis is often regarded as having made a major contribution to plant geography.

By the 1930s, vegetation studies were well established and much biogeography was merely regional description. *The British Islands and their Vegetation* (Tansley 1939) is, perhaps, the classic work of this period. There were also two important developments which were to have far-reaching impacts years later. The first was the publication of Tansley's paper in 1935 in which he coined the term *ecosystem*. This he defined as:

the whole system (in the sense of physics) including not only the organism-complex, but also the whole complex of physical factors forming what we would call the environment of the biome – the habitat factors in the widest sense.

A year later, in 1936, Marion Newbigin published her major work, *Plant and Animal Geography*. For several decades this book was to dominate the reading lists of biogeography courses taught in geography departments and was one of the first texts to have an ecological emphasis. Also written in the late 1930s was Ronald Good's global account of plant geography, *The Geography of the Flowering Plants* (which because of the outbreak of the Second World War was not published until 1947), and Ridley's *The Distribution of Plants Throughout the World* (1930).

The second development to have far-reaching effects on biogeographical studies in the British Isles was pollen analysis (palynology). During the last century it was discovered that pollen grains were preserved in their millions in peaty, acid environments. Not only did pollen provide a wealth of information on temporal changes in vegetation but could also provide information on its invasion and retreat from an area. In 1916, the Swedish geologist Lennart von Post put forward the idea of quantifying pollen information such that counts of various types of pollen were recorded against the stratigraphic position of the sample from which they were extracted. In this way, a time dimension was added to the study of vegetation.

In the mid-1930s the Irish Quaternary Research Committee invited the famous Danish palynologist Knud Jessen to visit Ireland. Professor Frank Mitchell, then a young student at Trinity College, Dublin, was his field assistant during Jessen's study of Irish bogs in 1934 and 1935. At about the time that Mitchell's formidable career started, Ireland's greatest naturalist, R. L. Praeger published *The Botanist in Ireland* which still remains a classic account of the flora of Ireland and has many perceptive comments on biogeography. Meanwhile, palaeoecological studies were started at Cambridge University by Professor Harry Godwin and his wife, and by 1940 Godwin had published the first tentative pollen-zonation for the post-glacial deposits in the British Isles. Godwin's studies, which inspired a generation of palynologists, were to culminate in the publication of *The History of the British Flora: a Factual Basis for Phytogeography* (1956).

The modern period

During the last 30 years or so there has been an enormous upsurge of interest in biogeography and it is difficult to characterize all the changes and developments which have occurred. Many important scientific bodies came into being in this post-war boom, higher education expanded tremendously, and major scientific discoveries such as plate tectonics belong to this period. Biogeographical studies of one kind or another are now taught in most universities and polytechnics.

Table 1.2 Fields of research activity of British biogeographers in the 1970s

1. Historical biogeography
   (a) Quaternary ecology, including limnology
   (b) early cultural biogeography
   (c) historical ecology
2. Contemporary ecosystems and their management
   (a) community description and dynamics
   (b) impact assessment
   (c) management, including recreation ecology
   (d) remote sensing
3. Biotic resources
   (a) general approaches
   (b) specific resources, e.g. mineral prospecting via biota
   (c) land-use ecology

Source: Simmons (1980)

Modern biogeography is very diverse, and different approaches are used by geographers, botanists, and geologists. Simmons (1980) suggests three major fields of research activity amongst members of the Biogeography Study Group of the Institute of British Geographers (see Table 1.2). But in a symposium on biogeography held at the British Museum and attended mostly by palaeontologists and biologists, Patterson (1981 and 1983) suggests that three very different schools of biogeography have emerged (see Table 1.3). Geographers have evidently become ecologists, and geologists – apart from the equilibrium theorists – are mostly not concerned with ecology. In the British Isles it has been left to the botanists and zoologists to emphasize mapping and the examination of spatial distributions as basic themes in biogeography; they seem to have little difficulty in distinguishing between ecology and biogeography. On the other hand, the distinction is sometimes blurred or simply not important for the geographer, who now has to cope with a plethora of methodological approaches – the study of spatial distributions and man–environment relationships being but two of many.

After the disruption of the Second World War a major boost to ecological and biogeographical studies in Britain was provided by the

Table 1.3 Schools of biogeography – the geological viewpoint

1. Wallace's methods – updated
   Invented by A. R. Wallace in his *Geographical Distribution of Animals* (1876). Originally the study of dispersal from centres of origin, coupled with evolution, over stable continents and oceans. Now updated to take into account plate tectonic theory.
2. Equilibrium methods
   These have grown out of MacArthur and Wilson's 1967 book, *The Theory of Island Biogeography*. Equilibrium theorists apply the mathematical models of population genetics and ecology to biogeography, and investigate the chances of dispersal, colonization, and extinction empirically, through studies of natural and man-made islands.
3. Vicariance biogeography
   Based on the work of Leon Croizat who believes that tectonic events have split whole biotas and that historical relationships between areas may be constructed through the systematic relationships of species. According to the vicariists, the distribution of life is governed not by the wanderings of plants and animals, but by earth history.

Source: Modified from Patterson (1981, 1983)

establishment of the Nature Conservancy in 1949, which was renamed the Nature Conservancy Council (NCC) in 1973. Throughout the 1950s and early 1960s the Nature Conservancy was directed by Max Nicholson, who led a very successful campaign to arouse a greater national interest in conservation and ecology. This was helped to a great degree by the publication in America of Rachel Carson's classic *Silent Spring* (1963), in which she painted a horrific account of the impact of pollution on the environment. The mechanisms of pollution processes could also be viewed in the context of ecosystems and Kenneth Mellanby's *Pesticides and Pollution*, published in 1967, is a thoughtful complement written by a scientist with a distinguished career at the Nature Conservancy. Geographers were particularly attracted by the ecosystems approach (Stoddart 1965) and most of the texts produced by them have a strong ecological emphasis. *Vegetation and Soils* (Eyre 1968), *The Ecology of Natural Resources* (Simmons 1974), *Basic Biogeography* (Pears 1985), and Tivy and O'Hare's *Man and the Ecosystem* (1980) exemplify this type of approach. Some, such as the very comprehensive *Geography of the Biosphere* by Furley and Newey (1983) also have detailed accounts of soils, and thus incorporate pedology into the corpus of biogeographical knowledge. The idea that the soil–vegetation complex was central to biogeography was certainly a popular one in the early 1960s. Two texts which did not follow the ecosystems approach were George's *Animal Geography* (1962) and Seddon's *Introduction to Biogeography* (1971). It is no coincidence that neither author is a geographer by training, Wilma George being a zoologist and Brian Seddon, although for several years in a geography department, is a botanist and now curator of natural history at Birmingham Museum.

In 1974, the *Journal of Biogeography* was launched under the editorship of David Watts, a biogeographer in the geography department at the University of Hull. In the same year, the Institute of British Geographers formed a Biogeography Study Group.

In 1963 the American ecologists MacArthur and Wilson published an influential paper which outlined their theory of island biogeography. Four years later, in 1967, they set out the theory in full in a monograph entitled: *The Theory of Island Biogeography*. This theory simply states that the number of species on an island is determined by a balance between immigration and extinction. Immigration rates will vary with the distance of the island from the mainland and extinction rates will vary with the size of the island. In spite of the enormous impact of the equilibrium theory in the USA it has only received scant attention by biogeographers in the British Isles, most notably by biologists at the University of York. It is puzzling why there has been so little attention given to the biogeography of the several hundred islands that make up the British Isles (Williamson 1981).

A second important development in the early post-war decades was the call, particularly by botanists, for better distributional data. In spite of the fact that there was a long tradition of biological recording and the use of Watson's vice-county recording scheme, there was some dissatisfaction, and in the early 1950s much effort was put into the establishment of grid-square data bases which could be easily managed by computer (see Chapter three). In the modern world of supercomputers and microchips with everything, it is all too easy to overlook the fact that at this time computers were very much in their infancy. Much of the credit for instigating the mapping schemes of the Botanical Society of the British Isles goes to A. R. Clapham and S. M. Walters, botanists at Sheffield and Cambridge Universities respectively who, with brilliant foresight, envisaged a world of computerized data bases and automated cartography. A major achievement of this initiative was the publication of *The Atlas of the British Flora* (Perring and Walters 1962). This one publication was to have an enormous impact on specialist groups and local organizations, many of which embarked on recording and mapping schemes of their own, and amateur naturalists were to play an important part in the collection of spatially referenced biological data. During the 1960s there was a great upsurge of interest in natural history and without the help of keen amateur recorders many county floras and atlases would not have been published. To give one example, the Royal Society for the Protection of Birds went from 8,000 members in 1959 to 300,000 in 1980 – many of whom were among the 10,000 or so observers who collected data for *The Atlas of Breeding Birds in Britain and Ireland* (Sharrock 1976)

One notable development was the establishment, in 1964, of the Biological Records Centre (BRC) at the Monks Wood Experimental Station. The BRC, now under the aegis of the Institute of Terrestrial Ecology's Environmental Information Centre, stores grid-square data on literally thousands of plant and animal species – all of which can be plotted in map form and are a major source of biogeographical data. In 1971, An Foras Forbartha established the Irish Biological Records Centre in Dublin with similar aims to the BRC. The importance of mapped biological data is a recurrent theme in *The Changing Flora and Fauna of Britain* (Hawksworth 1974) in which the biogeographical distributions of many taxa are discussed in depth.

Over the last 15 years or so the NCC have initiated a number of major habitat reviews. The first, *A Nature Conservation Review* (Ratcliffe 1977), classified the major terrestrial and freshwater habitats and described the biological features of more than 700 'key sites'. In 1987 a team of scientists at the NCC began work on *The Marine Nature Conservation Review* which will provide a description of British marine habitats and the distribution of species in the seas around Great Britain.

Closely linked to the growth of distribution studies was the growth in biogeographical studies of the Quaternary period and, in particular, studies of the late-glacial and post-glacial (from *c.* 15,000 years ago to the present day) faunal and floral history. Apart from the work of Godwin (1956) two other important textbooks related to these studies appeared in the decade 1950–60. In 1952, the entomologist B. P. Beirne published *The Origin and History of the British Fauna*, and, three years later, the botanist J. R. Matthews published *Origin and Distribution of the British Flora*, in the preface to which he acknowledged the pioneering contributions of G. F. Mitchell in Ireland and H. Godwin and his co-workers in Britain. Godwin went on to establish the sub-department of Quaternary Research in the Botany School at Cambridge which has become the pre-eminent centre for Quaternary palaeoecological research and training. Another significant contribution in this field has been made at the University of Birmingham geology department under the direction of F. W. Shotton, a contemporary of Godwin and Mitchell. Of note is the work of G. R. Coope on Quaternary insect faunas. At Imperial College (London University) M. P. Kerney has made a major contribution to our understanding of the Quaternary and present-day distributions of non-marine molluscs (Kerney 1976).

No account of the scope of British biogeography would be complete without a brief comment on biogeography and the theory of plate tectonics. Although the ideas about continental drift were originally put forward by Wegener as early as 1912 it was not until the late 1960s that the processes of sea-floor spreading and plate tectonics were understood. The story starts with the publication in Caracus of Leon Croizat's *Panbiogeography* (1958) in which he put forward the idea that, instead of species migrating from one area to another and thus fragmenting the range, disjunct distributions were to be seen as the result of tectonic movements, and that they represent the ranges of former taxa. Croizat's method was to link areas of endemism of similar taxa by a line or track on a map. The tracks for several unrelated taxa may coincide and form a generalized track. The global pattern of such tracks are statements about the former connections of either end of the tracks. Croizat did not interpret tracks as dispersal routes but as evidence of the fragmentation of formerly continuous ranges by the events surrounding plate movements. Such fragmented distributions are called vicariant distributions and can also result from geomorphological and climatic changes. The evidence for the order of fragmentation of a former land mass can be summarized in a branched, tree-like diagram called an area cladogram. Similarly, phyletic (or phylogenetic) cladograms can be constructed to illustrate the evolutionary relationships of taxa. Cladistic biogeography employs both methods to explain the historical evolution of present-day plant and animal distributions. Not surprisingly, cladistic biogeography is very much the domain of biologists and palaeontologists as the method relies heavily on a sound knowledge of taxonomy. Several zoologists and palaeontologists at the British Museum have written on cladistic biogeography and a summary of the methods can be found in *Cladistic Biogeography* (Humphries and Parenti 1986). A good general discussion of vicariance biogeography can be found in Brown and Gibson (1983).

Some remaining issues

In concluding this chapter I want to comment again on some of the questions raised earlier, and particularly the problem of scale and what organisms are suitable for study.

The question, 'what is an appropriate scale of study?' is made difficult because of the rather hazy boundary between ecology and biogeography. Fosberg (1976) attempts to circumscribe ecology as being essentially local in character whilst biogeography is regional, continental, or global. But in reality there are no natural breaks in the spatial hierarchy and it is probably not worthwhile trying to defend some arbitrary position. It is clear that the study of plant and animal distributions at the global, national, and even local scales – such as a county – can be accommodated within biogeography. It might upset some traditionalists, however, if it was argued that the distribution of nematodes in a single field was a legitimate part of the discipline. But perhaps the field is to the nematode what the globe is to a large mammal? And just as sea-floor spreading can give rise to vicariant distributions of mammals, cannot the construction of a motorway ultimately be a vicariant event to a soil invertebrate? Regardless of the geographical scale, the emphasis should be spatial rather than temporal, because that is what geography is all about. To my mind there is a tendency in some branches of biogeography, particularly Quaternary palaeoecology, for the one peat bog under study to be the 'centre of the world' and the spatial domain is subservient to temporal issues.

On a related matter perhaps it might be argued that different bio-geographical methods are applicable to different spatial and temporal scales of inquiry. Stoddart (1981) comments on this point and quotes Udvardy's three scales:

1 Secular scale – with spatial dimensions of about 100 km and time dimensions of about 100 years. At this scale, distributions are studied as biological phenomena subject to such factors as competition and dispersal abilities.
2 Millennial scale – covering several tens of thousands of years and scales up to 1,000 km in which climatic and geomorphic factors are important.
3 Phylogenetic scale – (or evolutionary time) where the time scale is of the order of 500 million years and the spatial extends to 40,000 km. At this scale, the ideas of plate tectonics and continental displacement are important.

The problems of scale partly arise through inadequate sampling. We know, for example, that aphid distributions over northern Europe fluctuate weekly and even daily in high summer (Taylor 1986). But dynamics of aphid distributions, say at the county level, can only be detected by daily monitoring at the appropriate scale.

The second question I want to examine briefly is, 'what organisms are suitable for biogeographical study?'. On the whole, biogeographers study relatively large plants and animals. These are better known and easier to observe, and, without doing anyone an injustice, it is probably true to say that many biogeographers are poor taxonomists. There is no logical reason to limit the scope of biogeography by the size or type of taxa. Legitimate biogeographical studies are possible with groups such as bacteria and viruses but, at this level, biogeography may impinge on plant and animal pathology, medical geography, agricultural geography, and even epidemiology, although I would prefer to say that they impinged on biogeography! After all, is there really any difference between a biogeographical study of a pathogenic virus which just happens to be carried along in the human bloodstream and is transmitted through social networks and, say, virus-vector soil nematodes whose present-day distribution in England and Wales is clearly related to the loamy soils south of the limit reached by the last glaciation?

Chapter two

The physical environment

Introduction

It has long been recognized that environmental factors such as soil type, temperature, salinity, and moisture are important in determining the distributions of plants and animals. For these reasons, some knowledge of the broad patterns in the physical environment of the British Isles is necessary. At this level of spatial resolution we can only attempt to relate nationally collected environmental data to biogeographical distributions down to, say, the county level. This is not to say that county boundaries actually coincide with environmental regions. The point is simply that in order to explain distributions *within* a local region we must obviously make use of local data.

Beyond this level of resolution it should be remembered that the boundaries of many environmental variables, such as temperature or rainfall, are only approximate since they are based on statistical averages. For example, the British Meteorological Service produces rainfall maps based on a 30-year period, and in what sense does a biogeographical distribution relate to an average, rather than, say, an extreme? Equally, we should realize that the definition of a biogeographical boundary is also only approximate. Distributions are often dynamic, and there is little point in worrying too much about minute details such as the thickness of the boundary line on the map, how any cartographic interpolation has been achieved, and whether or not all observations have been verified for a particular recording period.

The marine environment

Shelf seas

The coastal waters and seas around the British Isles are relatively shallow and rest on a wide continental shelf which slopes gently to the west and is in general devoid of any prominent topographical features. The margin of the shelf lies approximately at the 200 m isobath, and marks the boundary between the deep waters of the Atlantic Ocean and those spilling on to the shelf. At its nearest, the edge of the continental shelf is some 50 km west of Ireland and 120 km north of the Scottish mainland. The edge of the shelf is not only a major physical boundary, with the continental slope extending steeply down to the Atlantic Ocean floor several thousand metres below, but is also an important biogeographical one, with many taxa confined to one side or the other. For the most part, water depths on the shelf are less than 100 m, shallowing around the coasts and in the southern North Sea to 50 m or less (see Figure 2.1).

Figure 2.1 Major topographical features of the British Isles and surrounding seas.

Vertical zonation

Marine organisms can either float or swim, or they can live on, near, or within the sea bed. The entire sea-bed environment is called the *benthic division* and the water body forming the seas and oceans, the *pelagic division* (Figure 2.2). In the shallow waters over the continental shelf plant and animal life is richer and there is more variation in temperature and chemistry. For this reason the pelagic division is sub-divided into the *oceanic province* of the open ocean beyond the continental shelf, and the *neritic province* of shelf waters. In deep waters beyond the continental shelf, the oceanic province is further zoned on the basis of light and temperature characteristics into: (i) an *epipelagic zone* from the surface down to 200 m depth, in which there are sharp gradients of illumination and temperature; (ii) a *mesopelagic zone* from 200 m to 1,000 m depth which has a stable temperature and little illumination; (iii) a *bathypelagic zone* between 1,000 m and 4,000 m depth which is cold and completely dark; and (iv) an *abyssopelagic zone* below 4,000 m where water pressure is high, temperatures are low, and there is very little biological activity.

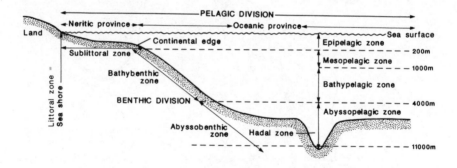

Figure 2.2 Cross section of marine environments. (Redrawn from Tait 1981)

The benthic division is also subdivided into three zones. The *littoral zone* includes the seashore and wave-splash region above the high water level. The *sub-littoral zone* extends from beyond the sea shore to the continental shelf, below which extends the *deep sea zone* which is sometimes divided into an upper bathybenthic and lower abyssobenthic zone. No deep-sea-zone environments occur in the British shelf seas.

Corresponding to the broad vertical zonation of the marine environment is a classification of marine organisms into three major groups. Organisms of the pelagic division comprise floating *plankton* which drift with the currents, and the *nekton* which are capable of travel independent of the

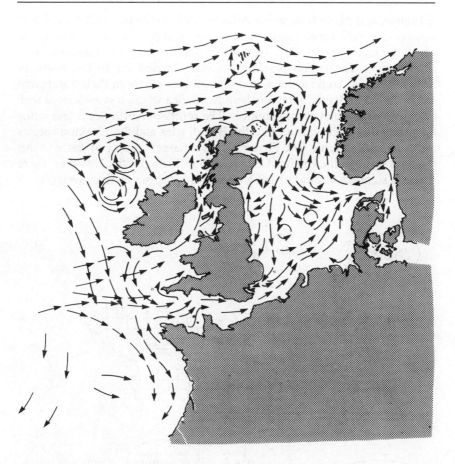

Figure 2.3 Residual surface currents in the north-east Atlantic. (Redrawn from Earll and Farnham 1983)

flow of water. Organisms of the benthic division be they sessile or burrowing, are collectively call *benthos*.

Water body characteristics

The waters of the neritic province are derived from a number of sources which can often be identified by characteristic plankton species. By far the largest influx of water over the shelf region comes from the North Atlantic Drift, and under strong flow conditions this water may reach the eastern end of the Channel and the Irish Sea. It is also carried round the north coast of Scotland into the northern part of the North Sea, for which it is the most important souce of water (see Figure 2.3). A second body of water is derived partly from deep Mediterranean water flowing over the sill of

Gibraltar, and partly from warm Atlantic water in the Bay of Biscay. On its way north it wells up and mixes with oceanic and coastal waters. The site at which the mixing takes place varies from year to year but exceptionally its characteristic *Lusitanian* plankton may be carried far to the north of Scotland. Cold Arctic water sometimes invades northern shelf waters and can be found at depth in the northern North Sea where it is associated with a cold-water population of plankton. The southern North Sea is less saline than the other shelf seas around the British Isles and the plankton species are dominantly neritic (see Figure 2.4). The largest supply of water to this region is through the English Channel and from the Baltic. However, there is also a significant input of fresh water from the major river systems.

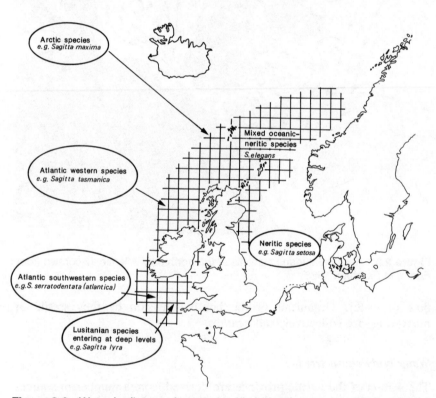

Figure 2.4 Water bodies as characterized by plankton species. (Redrawn from Tait 1981)

Two characteristics of water bodies which are of interest to the bio-geographer are salinity and temperature. The salinity of water is measured in parts per thousand and describes the total weight of dissolved salts present in a known volume of water. The North Atlantic Ocean water has a salinity of the order of 35 ‰, whereas most coastal waters have salinities one or two parts per thousand less. Maps of salinity levels show very clearly the penetration of tongues of Atlantic Ocean water into the north North Sea and the Western Approaches (see Figure 2.5).

Temperatures of the main bodies of water washing the continental shelf

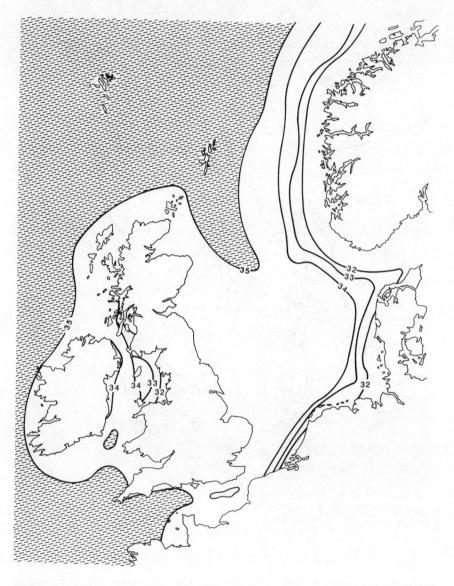

Figure 2.5 Mean surface salinity levels (‰). (Redrawn from Lee and Ramster 1981)

salinity of the order of 35.30, whereas surface coastal waters have salinities one or two parts per thousand less. Outside the shelf edge, show very clearly the penetration of upwelling Atlantic Ocean water into the Irish Sea, and the Western Approaches (see Figure 2.2).

Temperature of the main bodies of water washing the continental shelf

(a)

Figure 2.6 Mean summer temperature (°C) of (a) surface water; (b) bottom water. (Redrawn from Lee and Ramster 1981)

24

(b)

Figure 2.6 continued

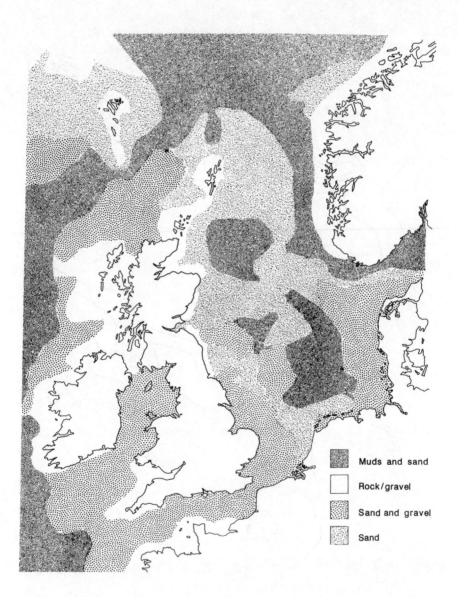

Figure 2.7 Major bottom deposits on the continental shelf. (Redrawn from Lee and Ramster 1981)

vary not only seasonally but also with depth. In winter here is virtually no difference in temperature between surface and bottom waters. Average temperatures are of the order of 5 °C to 6 °C and are lower than adjacent oceanic water by about 4 °C. The situation in summer is complicated by the fact that there is the development of a strong vertical temperature gradient, particularly in the northern North Sea basin. Surface-water temperatures show a simple latitudinal zonation locally modified in the shallow waters of the southern North Sea. Bottom temperatures vary much more sharply, and southern coastal waters may be as much as 10 °C warmer than more northerly locations (see Figure 2.6).

Sea-bed deposits

As with soils on land, the nature of the substrate into which organisms burrow or are attached is an important factor influencing species distributions. Several times during the last two million years much of the continental shelf and the islands rising from it have been covered by major ice-sheets which laid down glacial tills, sands, and gravels. As these ice-sheets wasted, and before world-wide sea levels rose, the glacial deposits were washed and eroded by river systems draining from the ice fronts and from land. In interglacial periods, waves and currents have continued to scour and re-arrange the glacial sediments and, in coastal zones, mix them with estuarine muds.

The general distribution of bottom deposits is shown in Figure 2.7, and can be related to the peak velocities of tidal currents found around the British Isles. The North Sea contains large areas of mud derived from the surrounding river systems, and sands and gravels derived mainly from glacial deposits. In the Irish Sea and Western Approaches the sands are mainly shell debris. Along the western and north-western parts of the continental shelf the sea bed is much more rocky, and reflects the fact that ice-sheets were essentially erosive in the north-west of the British Isles.

The land

The British Isles are an archipelago of some 400 islands most of which are small and lie close to the mainland of the two largest islands, Great Britain (or Britain – area 228,666 sq km) and Ireland (area 84,000 sq km). For the most part the islands are only separated from the mainlands by waters less than 20 m deep. A small number, including the Outer Hebrides, the Shetlands, the Scilly Isles, and the Isle of Man, are more distant and are surrounded by deeper water.

Both in Britain and Ireland the topography can be conveniently thought of as comprising either exposed uplands or more sheltered lowland plains. A major part of Ireland lies in the poorly drained Central Plain which is generally less than 150 m above sea level. This is surrounded on the north, west, and south by rolling uplands above 250 m, and several mountain ranges rising above 650 m. In Britain the major plains are confined to the east and south-east of the country. Much of Wales, northern England, and Scotland is composed of damp, exposed plateaux lying above 300 m OD and with large tracts, particularly in Scotland, above 700 m OD.

Climatic zonation

Of all the factors that influence the distribution of plants and animals on land, climate is the most important. In the British Isles two fundamental factors shape the climate. Of prime importance is the fact that the British Isles lie astride the tracts of frontal systems coming in from the Atlantic Ocean. It is these cyclonic systems that give rise to our maritime climates. Also important is the location of high mountains in the north and west, across which the major weather systems pass. Without these mountains the maritime influence of the westerly winds would be much more pronounced.

When considering the effect of climatic factors on plant and animal distribution, it should be borne in mind that climatic data are collected from instruments housed in Stevenson screens and are protected from insolation and wind. They are thus useful for building up a picture of regional climates but may not reflect the microclimates experienced by organisms. Sometimes the data may not even properly reflect the regional climate. This point is made rather nicely by the great Irish botanist, Lloyd-Praeger who wrote:

> The average rainfall at Kylemore in Connemara is given as 81.79 in. But the total rainfall conveys no indication of the number of damp, cloudy, drizzly and showers days that prevail in western Ireland which have much greater effect on plant life than deluges of rain lasting though a comparatively short period.

A last point to bear in mind is the paucity of good climatic data for the mountainous regions of the British Isles. In Britain there are only a handful of meteorological stations above 300 m OD, and our knowledge of upland climates comes mostly from poor extrapolations from lowland stations (Taylor 1976).

For the biogeographer, three climatic elements are important: first, the distribution of warmth in summer; second, the distribution of cold in winter; and, third, the distribution of precipitation.

Summer warmth

The distribution of July mean temperatures reduced to sea level is shown in Figure 2.8. Isotherms trend in a north-easterly direction and the British

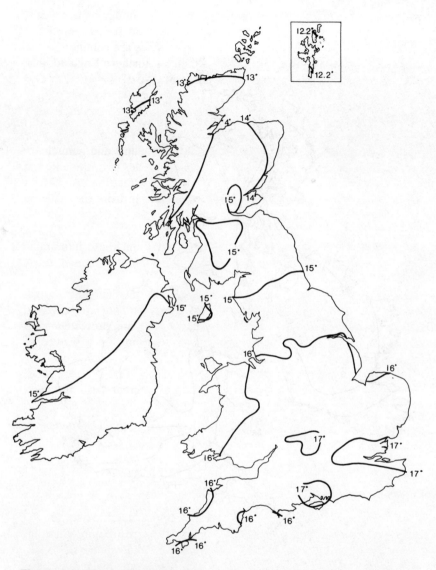

Figure 2.8 Mean July temperatures, 1941–71 (in °C), reduced to sea level.
(Redrawn from Tout 1976)

Isles is zoned into a cooler north and north-west and a warmer south and
south-east. South of a line through Liverpool to Hull temperatures are
above 16 °C. In the far north of Scotland July temperatures are 3–4 °C
lower, the greater day length failing to make up for the general cloudiness
and exposure. The isotherms shown in Figure 2.8 do not, of course, reflect
summer temperatures in the mountains and to obtain some approximate

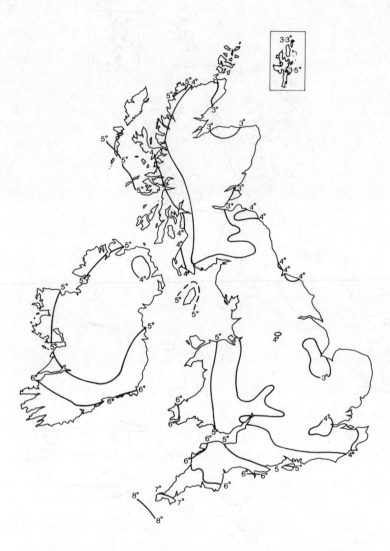

Figure 2.9 Mean January temperatures, 1941–71 (in °C), reduced to sea level. (Redrawn from Tout 1976)

estimates for these a lapse rate correction of −0.6 °C per 100 m rise in elevation can be applied. In reality this means that air temperatures on the high mountains of Scotland are unlikely to be much more than 10 °C and the sensible temperatures, as far as plants and animals are concerned, even lower because of the general windiness at such altitudes (Taylor 1976).

Winter cold

The January mean temperatures, reduced to sea level, are shown in Figure 2.9. The general north–south trend in winter isotherms over the British Isles is clearly influenced by weather systems coming in from the Atlantic, the isotherms bending inland over southern Ireland as a result of maritime influences. Over Ireland and much of south-west England and Wales, January temperatures are in excess of 5 °C, and in the extreme south-west of England reach as high as 8 °C. In these southerly locations, plant growth is possible throughout the year. In the uplands of Britain winter conditions can be severe. At high altitudes winds become more persistent and the wind force stronger. On Ben Nevis (1,343 m OD) Taylor (1976) notes that over a period of 13 years there were on average 261 gales of force greater than 80 kph. At such altitudes, both wind stress and frost inhibit plant growth. In addition, freeze–thaw cycles induce soil disturbance, making rooting conditions difficult.

Precipitation

The general distribution of precipitation in the British Isles is shown in Figure 2.10. At sea level, precipitation shows a north–south trend but there are also strong orographic controls and, as a result, precipitation totals vary considerably. In the mountains of Kerry and Donegal, Snowdonia, the Lake District and north-west Scotland annual totals are in excess of 2,400 mm. This is in stark contrast to much of eastern England where the total is less than 800 mm.

Above 700 m OD winter precipitation is frequently of snow, particilarly in Scotland and northern England. But in spite of the severity of winter conditions there are no permanent snow-fields, although one or two isolated snow patches linger from year to year in north-facing gullies on Ben Nevis and in the Cairngorms. On average, Ben Nevis is usually free of snow-cover for some 75 days a year. The duration of snow-lie in early summer probably has a good deal of influence on the distribution of montane plants and animals, but has not been studied in the British Isles.

Length of the growing season

Although plants may have particular temperature preferences, very little vegetative growth is possible below 6 °C. When the daily mean temperature rises to this level, the growing season is said to have started, and it ends during the autumn when the average daily temperature fails to reach that threshold. The length of the growing season is very simple to calculate and is a useful statistic which brings out a number of points (see Figure 2.11). Most of Ireland, Wales, and southern England has a vegetative

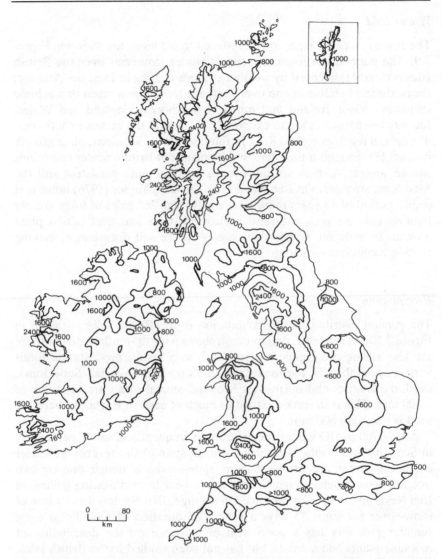

Figure 2.10 Mean annual rainfall (in mm) 1931–60. (Redrawn from Tout 1976)

period in excess of 240 days, in contrast to the cold areas of the Pennines, the Southern Uplands, and Highlands of Scotland. Where the isolines are close together, as in northern Britain, they indicate a rapid reduction in the growing season and this is very noticeable as the uplands are approached. The rate of reduction varies considerably. In Scotland the average decrease in days per 30.5 m (100 feet) is 4.8 as compared with 10.8 for south-west England and south-west Ireland.

Biogeographers often map temperature data reduced to sea level in

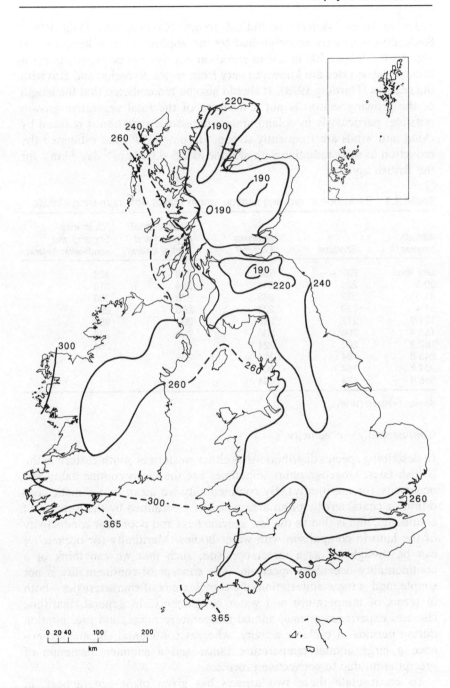

Figure 2.11 Actual lengths of the growing season (or vegetative period), based on annual mean temperatures. (Redrawn and modified from Fairbairn 1968)

order to reveal general latitudinal trends (Conolly and Dahl 1970). Reduction is usually accomplished by the application of a lapse rate of about −0.6 °C per 100 m rise in elevation but this will be slightly in error because lapse rates are known to vary from region to region and also with the seasons (Harding 1978). It should also be remembered that the length of the growing season is not a measure of the total vegetative growth possible, particularly in upland areas where solar radiation is reduced by cloud and winds are frequently strong. Harding (1979) has estimated the reduction in solar radiation with altitude as 2.5–3.0 MJ m^{-2} day^{-1} km^{-1} for the British uplands.

Table 2.1 Reduction in growing season length (days) with increasing altitude

Altitude (metres)	Scotland	Northern England	South-central England and central Ireland	South-west England and south-west Ireland
Sea level	232	252	292	328
30.5	228	248	284	318
61.0	222	243	276	308
91.4	220	238	268	296
121.9	212	234	260	286
152.4	208	230	250	274
182.9	202	224	242	264
243.8	194	214	224	
304.8	184	204		
365.8	174	194		

Source: Fairbairn (1968)

Continentality – oceanicity

In describing species distributions as either western or south-eastern in the British Isles, biogeographers sometimes use the terms continentality and oceanicity (or maritimity). By continentality we mean that climatic condition or characteristic of an area which is determined by the location of continents, and is due to the low specific heat and poor heat conductivity of the land in comparison with water bodies. Maritimity (or oceanicity) can be thought of as a direct opposite, such that we can think of a continentality–oceanicity spectrum. The concept of continentality is not simple, and is the manifestation of a complex mix of characteristics – both in terms of temperature and water components. In general, maritime climates experience a small annual temperature range, and precipitation during periods of cyclonic activity, whereas continental climates experience a large annual temperature range and a summer maximum of precipitation due to convective processes.

To encapsulate these two aspects has given plant geographers, in particular, much food for thought and they have devised a great many indices (Tuhkanen 1980). One common index based on the thermal

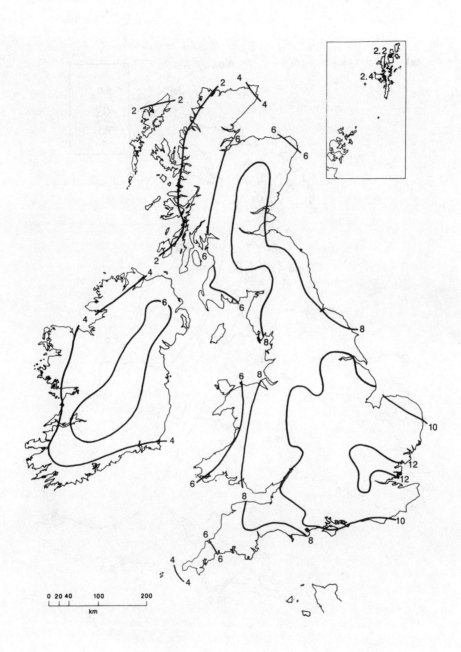

Figure 2.12 Conrad's continentality index, 1941–70. (Redrawn from Tout 1976)

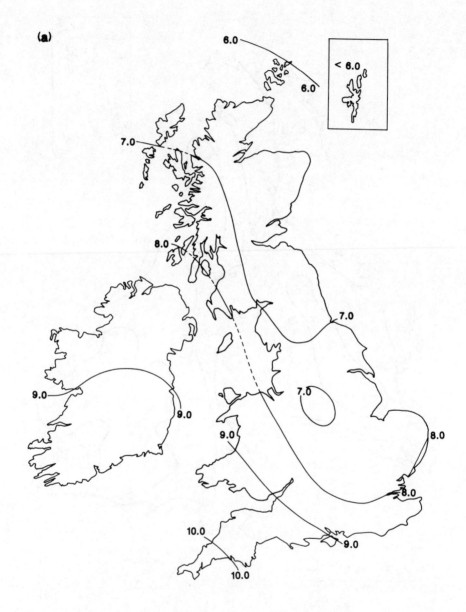

Figure 2.13 Average daily totals of global solar radiation (MJ m^{-2}) (a) March; (b) June; (c) September; (d) December. (Redrawn from Collingbourne 1976)

(b)

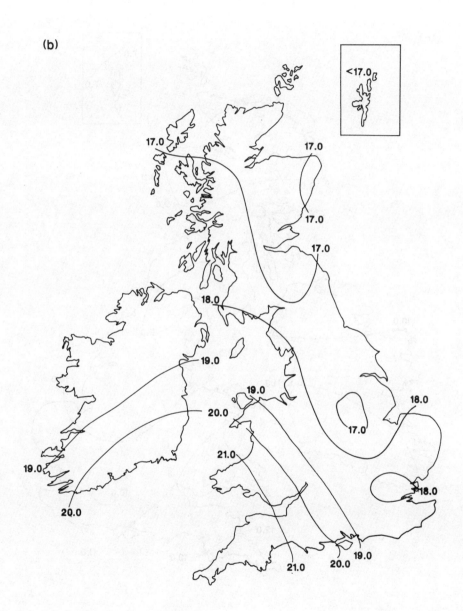

Figure 2.13 continued

(c)

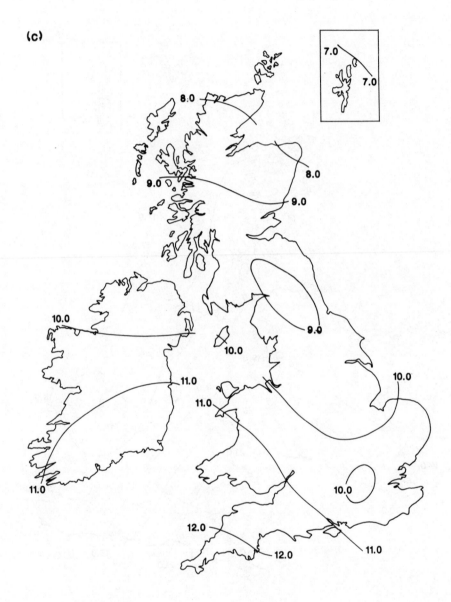

Figure 2.13 continued

(d)

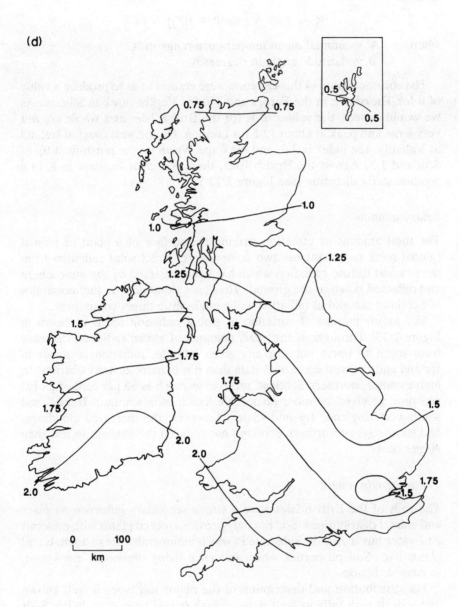

Figure 2.13 continued

component only is *Conrad's continentality index* (K) which is calculated as follows:

$$K = (1.7 \, A \, / \, \sin(\theta + 10°)) - 14$$

where   A = annual mean temperature range in °C
           θ = latitude angle in degrees.

The constants used in this equation were chosen so as to produce a value of 0 for Thorshavn in the Faroes and 100 for Verkhoyansk in Siberia. As we would expect, the values of K for the British Isles as a whole are not very large and peak at about 12.5 for London. On the west coast of Ireland at Valentia, the index is 2.6, and for Cape Wrath on the north-west tip of Scotland 1.3. Across the British Isles, there is a clear increase in K in a south-easterly direction (see Figure 2.12).

Solar radiation

The total amount of radiation striking the surface of a plant or animal (global solar radiation) has two components: direct solar radiation from the sun and diffuse radiation which has been scattered by the atmosphere and reflected down to the ground. Thus on a cloudy day in the mountains of Scotland the global radiation will contain little direct component.

The major pattern of variation in global radiation totals is shown in Figure 2.13. Throughout the year, amounts of global radiation decrease from south to north but, for any given latitude, radiation is higher in Ireland and the west coast of Britain than it is in more easterly districts. In high summer, northern Scotland receives as much as 85 per cent of global radiation received by more favourable locations in southern Ireland and south-west England. By mid-winter, however, the increased cloudiness and short days of northern Scotland has reduced the amount to less than 40 per cent.

The soil environment

The soils of the British Isles have a strong secondary influence on plant and animal distributions. Soil not only provides rooted plants with minerals and water but is also the substrate in which innumerable small animals and plants live. Soil properties which influence living organisms are known as *edaphic* factors.

The distribution and description of the major soil types is well known and is dealt with fully in *Soils in the British Isles* (Curtis *et al*. 1976). Soils are the product of a number of interrelated factors – the most important of which are climate, parent material, vegetation and associated organisms, relief, and time. In Figure 2.14 the general occurence of the major soil types can be seen broadly to correspond with the known variations in

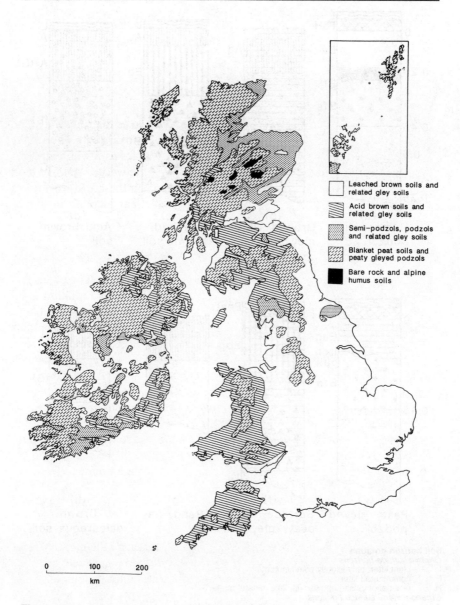

Figure 2.14 Major soil regions of the British Isles.

geology and climate. Well-drained calcareous soils are most common on the young sediments in the lowlands; poorly drained, acid soils predominate in the highlands and wetter western areas.

Soils are differentiated one from another by their *profile* characteristics, the profile being the arrangement of various layers as seen in vertical

41

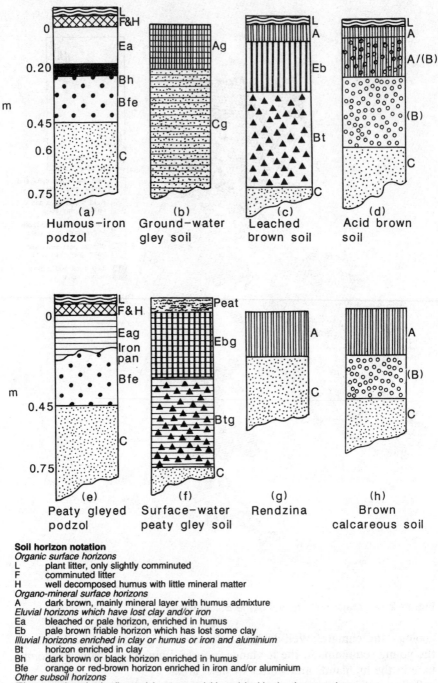

(a) Humous–iron podzol
(b) Ground–water gley soil
(c) Leached brown soil
(d) Acid brown soil
(e) Peaty gleyed podzol
(f) Surface–water peaty gley soil
(g) Rendzina
(h) Brown calcareous soil

Soil horizon notation
Organic surface horizons
L   plant litter, only slightly comminuted
F   comminuted litter
H   well decomposed humus with little mineral matter
Organo-mineral surface horizons
A   dark brown, mainly mineral layer with humus admixture
Eluvial horizons which have lost clay and/or iron
Ea   bleached or pale horizon, enriched in humus
Eb   pale brown friable horizon which has lost some clay
Illuvial horizons enriched in clay or humus or iron and aluminium
Bt   horizon enriched in clay
Bh   dark brown or black horizon enriched in humus
Bfe   orange or red-brown horizon enriched in iron and/or aluminium
Other subsoil horizons
(B)   weathered subsoil material, not appreciably enriched in clay, humus or iron,
       distinguished redrawn from overlying and underlying horizons by colour or structure or
       both

section (see Figure 2.15). The layers, or horizons, develop due to the *translocation* of soluble and insoluble constituents mainly from higher to lower positions within the profile. Other things being equal, horizon development is strong when there is abundant rainfall and its infiltration into the soil is not impeded. The main causes of low infiltration rates are clayey textures and platy structures. Soil *texture* refers to the size of the mineral particles in the fine earth fraction which comprises all material which is less than 2 mm in diameter. By convention, the size classification is as follows: sand (2–0.02 mm), silt (0.02–0.002 mm), and clay (less than

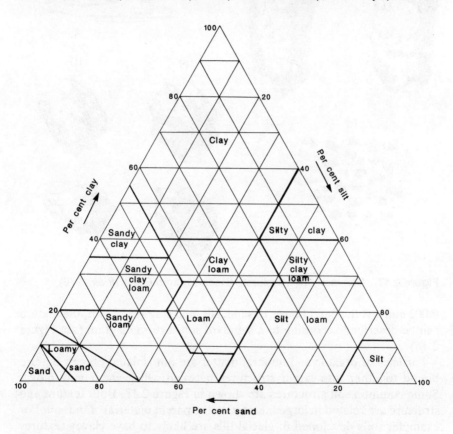

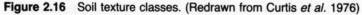

Per cent sand

Figure 2.16 Soil texture classes. (Redrawn from Curtis *et al.* 1976)

Figure 2.15 Profiles of common British soil types. (Redrawn from Burnham and Mackney 1964)

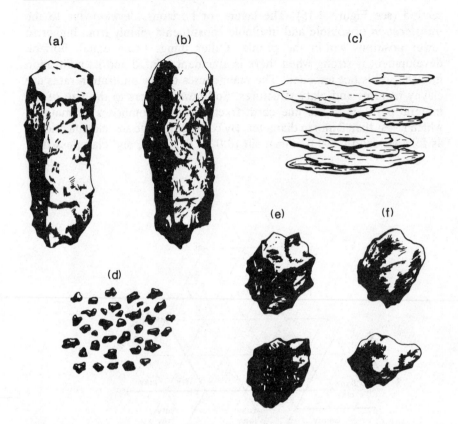

(a)   (b)   (c)

(e)   (f)

(d)

Figure 2.17 Common soil structures. (Redrawn from Curtis *et al.* 1976)

0.002 mm). If the percentages of sand, silt, and clay are known, the texture can be described according to a standard soil-texture diagram (see Figure 2.16).

Particles of organic and mineral matter are not isolated objects but are bound together into larger structures between which water may move. Some common soil structures are shown in Figure 2.17. Both texture and structure are related in large measure to the parent material of the soil. For example, soils developed in glacial tills are likely to have clayey textures and blocky structures and as a result horizon development is usually poor. In contrast, soils developed from friable sandstones are likely to be well drained and have good horizon definition.

If precipitation is effective, soluble cations, such as calcium, magnesium, potassium and sodium are *leached* down the profile, leaving behind immobile aluminium and iron oxides and a clay-mineral complex dominated by hydrogen ions. Under certain circumstances, aluminium and iron cations may be leached from their immobile oxides by a process called

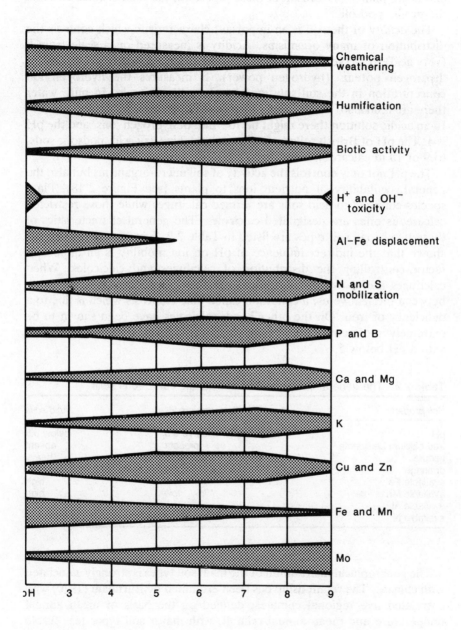

Figure 2.18 The influence of soil pH on soil formation, mobilization, availability of soil nutrients, and the conditions for life in the soil. (Redrawn from Larcher 1975)

cheluviation caused by chemical agents released from the litter of conifers and heath plants. As a result of these processes, the soil becomes acid and the profile podsolic.

The *acidity* of the soil is an important characteristic which controls the distribution of many organisms. Acidity is measured on a scale from 0 (very acid) to 14 (very alkaline). This is the pH scale, where pH stands for 'hydrogen potenz' (hydrogen power). It measures the hydrogen ion concentration in the soil solution on a logarithmic scale. In pure water there are 0.0000001 or 10^{-7} mol of hydrogen ions per litre and the pH is 7. In an acidic solution there might be 10^{-4} mol of hydrogen ions, and the pH is 4. The pH of British soils varies from about 3 in very acid, podsolic soils, to 9 or 10 in calcareous rendzinas.

The pH not only controls the activity of soil micro-organisms but also the general availability of nutrient ions to plants (see Figure 2.18). Plant species confined to acid soils are termed calcifuges while those restricted calcareous ones are designated calcicoles. The general characteristics of these two broad soil types are listed in Table 2.2. Experimental work has shown that the indirect influence of pH on ion mobility is an important factor controlling the distribution of calcifuges and calcicoles. When calcifuges are grown in calcareous soils they often develop a yellowing between the leaf veins, a condition called lime chlorosis which is due to a deficiency of iron. On the other hand, calcicoles have been shown to be extremely intolerant of aluminium which increases in concentration in soil with a pH below 5.

Table 2.2 The general characteristics of acid and calcareous soils

Soil property	Calcareous soils	Acid soils
pH	above 6.5	below 6.5
free calcium carbonate	present	absent
texture	light	heavy
drainage	good	poor
available Fe	low	high
available Mn	low	high
available Al	low	high
available N	high	low

The geographical distribution of the main soil types is strongly associated with climate. The relationship has been examined by Burnham (1970) who correlated five regional climates, defined on the basis of mean annual temperature and mean annual rainfall, with major soil types (see Table 2.3). Leached brown soils and brown podsolic soils (semi-podsols) are typical of much of lowland England where soils often dry out in summer. In highland Britain and over much of Ireland the soils are wet throughout much of the year and are subject to gleying. In such soils gaseous diffusion is poor, soil pores remain filled with water for much of the time and the soil profile takes on a mottled appearance due to the oxidation state of the iron present. High rainfall and poor drainage may also prevent the oxidation and breakdown of plant material which accumulates at the surface of the soil to form a peaty organic horizon. Over much of western Ireland and on the rolling upland surfaces of Britain peaty organic horizons have thickened to form blanket peat.

Table 2.3 Relationship between soil types and climatic regimes in Britain

Climatic region	Mean annual temperature (°C)	Mean annual rainfall (mm)	Characteristic soils
warm, dry	over 8.3	under 1000	leached brown soil
cold, dry	4.0–8.3		semi-podsols/podsol
warm, wet	over 8.3	over 1000	acid brown soil
cold, wet	4.0–8.3		peaty gley podsol/ blanket peat
very cold, wet	under 4.0		alpine humus soil

Source: Burnham (1970)

Chapter three

Recording species distributions

Introduction

As we saw in Chapter one, British biogeographers were among the first to recognize the importance of maps of species distributions. Of course, map production has come a long way from the days of paint-pot and pen, and nowadays much effort is being put into the production of distribution maps by computers. But in spite of these cartographic developments the actual, detailed distribution of many plants and animals is still poorly known. For this reason alone, mapping species distributions continues to play an important role in biogeography.

To the trained observer maps ask questions and sometimes provide answers. Figure 3.1 shows the contrasting distributions of two species of woodlice. *Ligidium hypnorum* is apparently confined to deciduous woodland and fens in south-east England, while *Porcellionides cingendus* is mainly found in south-west Ireland, where it is associated with open grassland and scrub, particularly around the coast. What sort of questions might we want to ask about these two distributions? In the case of *Ligidium* we might wonder why it does not occur in the south-west of England in spite of suitable habitats, and indeed, why it does not occur in the southern part of Ireland? Perhaps it used to occur in these areas but has died out, or perhaps there simply has not been time for it to disperse into these areas since conditions became favourable at the end of the last glacial stage, *c*.10,000 years ago.

The south-western distribution of *Porcellionides cingendus* suggests that it prefers mild, temperate climates, and that winter temperature might be a limiting factor. But if this is so, it also raises the question as to how its distribution evolved after the ice age. If it is so sensitive to winter cold then it is hardly likely to have survived the rigours of glacial and periglacial climates in southern Ireland. But if it migrated overland from northern France in the post-glacial it would have had to do so before the rise in sea level that was taking place flooded the English Channel and Irish Sea and thereby severed its routeway. Moreover, it is surprising that *Porcellionides*

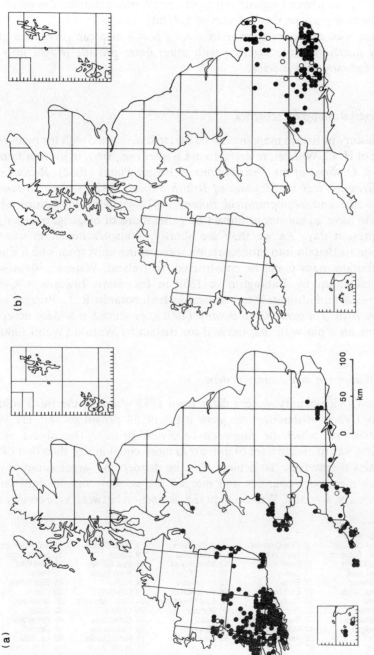

Figure 3.1 Distribution of two woodlice species using 10 km grid square mapping units: (a) *Porcellionides cingendus*; (b) *Ligidium hypnorum*. Open circles represent squares where there is only a record before 1970. (Redrawn from Harding and Sutton 1985)

is not more widely distributed in southern England. Are the population clusters in southern England relicts of a much wider distribution which has fragmented perhaps through loss of habitat?

These two examples show quite clearly how a map can generate a great many questions and, together with other data, provide insight into the biogeography of a species.

Terrestrial mapping schemes

The history of British mapping can be said to have started with the pioneering work of H. C. Watson, referred to in Chapter one, who, in his four famous books, *Outlines of the Distributions of British Plants* (1832), *Remarks on the Geographical Distribution of British Plants* (1835), *Cybele Britannica* (1847–59), and *Topographical Botany* (1873–4), laid foundations which were to have an enormous impact on generations of biogeographers up to the present day. As we shall see shortly, Watson's invention was the division of Britain into biogeographical recording units from which simple distribution maps could be constructed. In Ireland, Watson's ideas were first taken up by Babington in 1859 in his 'Hints towards a *Cybele Hibernica*' and ultimately by the famous Irish botanist R. L. Praeger who, in his *Irish Topographical Botany* (1901), produced a widely accepted scheme on a par with that devised for Britain by Watson (Webb 1980).

The Watsonian vice-county system

Watson's vice-county system dates from 1852 when, in the third volume of his *Cybele Britannica*, he gave a list of 38 provinces and 112 vice-counties into which he proposed to divide Britain. The object of his division was to create a set of unit areas more equal in size than that of the counties themselves. To achieve this, he divided the larger counties into two or more vice-counties and merged the smaller counties into larger neighbours. Initially, Watson kept the distinction between vice-county and

Great Britain

1 West Cornwall	2 East Cornwall	3 South Devon	4 North Devon	5 South Somerset
6 North Somerset	7 North Wilts	8 South Wilts	9 Dorset	10 Isle of Wight
11 South Hants	12 North Hants	13 West Sussex	14 East Sussex	15 East Kent
16 West Kent	17 Surrey	18 South Essex	19 North Essex	20 Herts
21 Middlesex	22 Berks	23 Oxfordshire	24 Bucks	25 East Suffolk
26 West Suffolk	27 East Norfolk	28 West Norfolk	29 Cambridge	30 Bedford
31 Hunts	32 Northampton	33 East Gloucester	34 West Gloucester	35 Monmouth
36 Hereford	37 Worcester	38 Warwick	39 Stafford	40 Salop
41 Glamorgan	42 Brecon	43 Radnor	44 Carmarthen	45 Pembroke
46 Cardigan	47 Montgomery	48 Merioneth	49 Caernarvon	50 Denbigh
51 Flint	52 Anglesey	53 South Lincoln	54 North Lincoln	55 Leicester
56 Notts	57 Derby	58 Chester	59 South Lancs	60 West Lancs
61 South-east Yorks	62 North-east Yorks	63 South-west Yorks	64 Mid-west Yorks	65 North-west Yorks
66 Durham	67 Northumberland South	68 Cheviotland	69 Westmorland	70 Cumberland

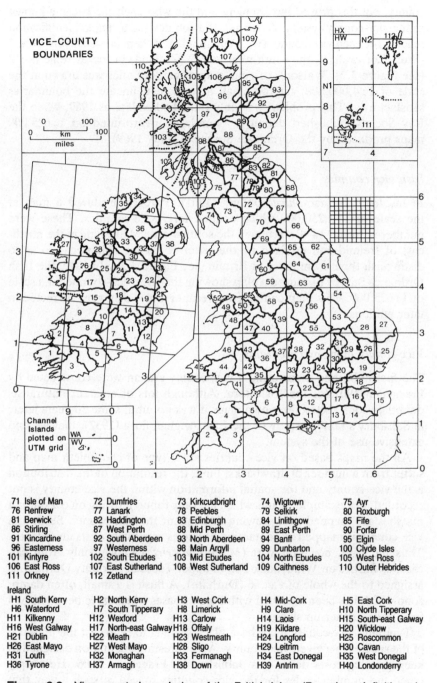

71 Isle of Man   72 Dumfries   73 Kirkcudbright   74 Wigtown   75 Ayr
76 Renfrew   77 Lanark   78 Peebles   79 Selkirk   80 Roxburgh
81 Berwick   82 Haddington   83 Edinburgh   84 Linlithgow   85 Fife
86 Stirling   87 West Perth   88 Mid Perth   89 East Perth   90 Forfar
91 Kincardine   92 South Aberdeen 93 North Aberdeen   94 Banff   95 Elgin
96 Easterness   97 Westerness   98 Main Argyll   99 Dunbarton   100 Clyde Isles
101 Kintyre   102 South Ebudes   103 Mid Ebudes   104 North Ebudes   105 West Ross
106 East Ross   107 East Sutherland 108 West Sutherland 109 Caithness   110 Outer Hebrides
111 Orkney   112 Zetland

Ireland
H1 South Kerry   H2 North Kerry   H3 West Cork   H4 Mid-Cork   H5 East Cork
H6 Waterford   H7 South Tipperary H8 Limerick   H9 Clare   H10 North Tipperary
H11 Kilkenny   H12 Wexford   H13 Carlow   H14 Laois   H15 South-east Galway
H16 West Galway   H17 North-east Galway H18 Offaly   H19 Kildare   H20 Wicklow
H21 Dublin   H22 Meath   H23 Westmeath   H24 Longford   H25 Roscommon
H26 East Mayo   H27 West Mayo   H28 Sligo   H29 Leitrim   H30 Cavan
H31 Louth   H32 Monaghan   H33 Fermanagh   H34 East Donegal   H35 West Donegal
H36 Tyrone   H37 Armagh   H38 Down   H39 Antrim   H40 Londonderry

Figure 3.2 Vice-county boundaries of the British Isles. (Based on definitions in Dandy 1969 and Webb 1980)

county, but this proved inconvenient and the whole scheme became known as the vice-county system. As an aid, Watson devised a simple identification scheme for each vice-county (v.c.) and numbered them consecutively from 1 (West Cornwall) in the south-west to 112 (Zetland) in the extreme north (see Figure 3.2). Watson's original map of vice-counties was drawn at the scale of 1:4,000,000, far too small a scale to delineate the boundaries satisfactorily. This problem was only remedied as late as 1969, when the Ray Society published definitive boundaries overprinted on 1:625,000 maps produced by the Ordnance Survey (Dandy 1969).

Irish vice-counties

In his *Irish Topographical Botany* (1901), Praeger produced a map at the scale of 1:1,250,000 which defined forty vice-counties. These were numbered from 1 to 40 starting in the south-west and finishing in the north-east of Ireland and prefixed by the letter H (H for Hibernica) so as to distinguish them from those of Britain (see Figure 3.2). In 1949, the Irish Ordnance Survey produced a map showing the Irish vice-counties at a scale of 1:625,000. This map contained several errors which were corrected in later editions.

Vice-county maps and recording

The first published maps using a vice-county system were constructed by Praeger in 1902 for the Irish flora. Although lots of amateur naturalists doubtless constructed their own maps it was not until some 30 years later, in Salisbury's famous study of the East Anglian flora (1932), that we find extensive use of the system.

All the maps based on vice-counties are a type of choropleth map and suffer from a number of drawbacks. First, the resolution of the distribution is the vice-county and the spatial information within the vice-county is not recorded. For example, plants which are only found growing on coastal salt marshes are mapped as if growing throughout the vice-county. Second, a vice-county is mapped as containing a species regardless of its abundance. Thus the rare, bog sandwort (*Minuartia stricta*) known only from cal-careous flushes on Widdybank Fell in upper Teesdale, County Durham, is assigned to the whole of v.c. 66 (Durham). A flush is a small, often damp, zone which has been enriched with lime or bases which have been leached from soils further up slope.

Typically, vice-county maps have black shading to represent the presence of the species in a given vice-county. An interesting adaptation designed to keep printing costs down was published by Praeger in 1906. He simply typed the number codes for the vice-counties spaced in such a way that they gave a reasonable representation of the shape of Ireland. Vice-

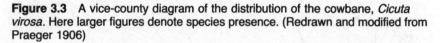

```
                              –  –
                     35 34 40  –  39
                     –  –  – 36  –  –  –
                     –  33  – 37 38  –
           –  27  – 28 29 30 32  –  –
              – 26  – 25 24  – 31
           – 16  – 17  –  – 23 22  –
              –  15  – 18  – 19 21
              –  9  – 10  – 14  – 20
              –  8  –  7  – 11 13
        –  –  2  –  –  –  –  – 12
        1  –  –  4  5  6
           –  –  3  –
```

Figure 3.3 A vice-county diagram of the distribution of the cowbane, *Cicuta virosa*. Here larger figures denote species presence. (Redrawn and modified from Praeger 1906)

counties containing the species were printed in heavy type. His invention is shown in Figure 3.3, which illustrates the distribution of cowbane (*Cicuta virosa*), a poisonous plant belonging to the carrot family. Other variants on simple choroplething were used effectively by Salisbury to illustrate the spread of immigrant species by grouping vice-counties in terms of the date of a species arrival, and applying different shading patterns to each group (see Figure 3.4).

There are also practical difficulties associated with the vice-county schemes. Both in Britain and Ireland, the late nineteenth century saw a number of Local Government Acts which made changes to several county boundaries and also created County Boroughs. For the vice-county system to be a practical recording system, the boundaries have to be stable. Plant and animal records cannot simply be transferred into a new geographical unit, because they were often never recorded with any more precision than their occurrence in a particular vice-county. In order not to lose the

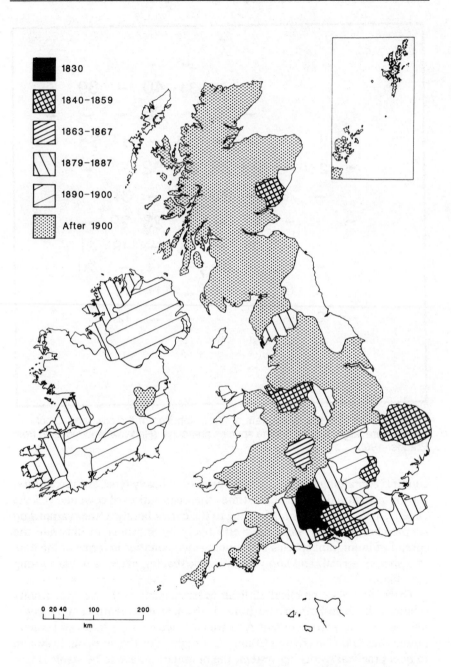

Figure 3.4 A vice-county choropleth map designed by Salisbury to show the date of first occurrence of the monkey-flower (*Mimulus guttatus*), an introduced species. (Redrawn and modified from Salisbury 1932)

enormous body of information that had been built up, the professional bodies handling the recorded data decided to keep to the original boundaries as defined by Watson and Praeger.

A second problem with the vice-county scheme relates to imprecise dividing lines. Where a county was divided into two or more vice-counties, the dividing line was usually chosen to follow a feature such as a river or road. In some cases, however, the definition was vague to say the least. For example, Watson defines v.c. 3 (South Devon) as:

Devonshire south of an imaginary line adapted to the water-shed; commencing at the Tamar, about mid-way between Tavistock and Launceston; passing over the ridge of Dartmoor, and joining the Western Canal at Tiverton

In Britain, difficult cases such as this were the subject of deliberation by a sub-committee of the Systematics Association, which was charged with the job of defining the boundary more precisely, and it is these boundaries which appear on the maps produced by the Ray Society (Dandy 1969). Given the scale of these maps and the nature of the Dartmoor watershed, one could be forgiven for thinking that one was in v.c. 3 rather than v.c. 4, especially on a foggy day!

A detailed definition of the Irish vice-county boundaries is given by Webb (1980), who writes with obvious enthusiasm about this method of recording. One difficulty with the vice-county system as applied to Ireland, at least for fresh-water organisms, is that so many boundaries follow rivers. Webb suggests that a single record will in many cases either be ambiguous or will give rise to records for two vice-counties. To overcome this problem, it was suggested that the basic recording unit should be the river basin. As Webb notes, however, this merely replaces the ambiguity for river organisms with a similar one for those of the mountains. Furthermore, Ireland is the wrong shape for such a division, since the Shannon basin is far too large and many coastal river basins too small.

Although there are undoubted disadvantages with vice-county recording schemes, we cannot ignore the fact that there is a vast body of data described in this format. Furthermore, the numbering scheme is such that it is very easy to picture a distribution from a list of the vice-county numbers because of the inherent geography in their designation. Problems of precision regarding boundary definitions do not go away by replacing wiggly lines by straight grid lines as we shall see.

Grid mapping

Watson was well aware of some of the problems associated with mapping species distributions in vice-counties, and in particular he realized that if the recording units could be made sufficiently small then some measure of the general abundance could be gleaned from the map. Unfortunately, during Watson's time, maps were not available for this development.

At the local level, some attempt was made to show species locations more precisely by the production of simple dot maps, the first of which for a British plant was published by Good in 1936. Each dot was of uniform size and represented a site at which the plant could be located. Such maps can be extremely helpful in understanding local distributions. Figure 3.5 shows such a map for the distribution of the hoary plantain (*Plantago media*) in Dorset. When this map is examined in conjunction with a geological map the hoary plantain is seen to be related to lime-rich strata and their associated calcareous soils.

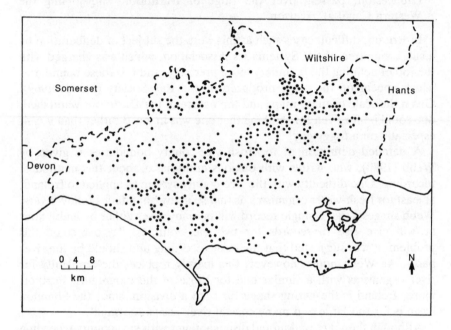

Figure 3.5 A dot map of the calcicole, hoary plantain (*Plantago media*). (Redrawn and modified from Good 1948)

The clear advantage of dot maps over vice-county maps did not go unnoticed. But even by the early 1950s, there simply was not enough detailed locational information to produce dot maps for anything other than rare species. For many parts of Britain no data were available other than that a species does or does not occur in a particular vice-county.

To some degree, the lack of precise locational data was not the fault of the recorders, but a reflection on the development of the British and Irish mapping systems. At this point we should be clear about the distinction between the graticule of the map projection and the map grid we use when citing a grid reference. The graticule is the network of meridians of longitude and parallels of latitude that are shown on maps and it is these which are drawn when trying to depict the transformation from the three-dimensional globe to the two-dimensional map. A grid, on the other hand, is an entirely arbitrary network of intersecting lines drawn parallel, and at right angles, to the central meridian, so as to enable the position of a place to be defined more simply than by latitude or longitude, and also to provide a single reference system for the whole country.

Nowadays, the map projection for both Irish and British maps is the Modified Transverse Mercator. This has an origin at latitude 53° 30' north and longitude 8° west of Greenwich for the Irish projection, and longitude 2° west and latitude 49° north for the British. On British maps, grid systems were used on military maps during the First and Second World Wars, but the present-day National Grid dates from a recommendation by the 1938 Davidson Committee. The relationship between the Transverse Mercator graticule and the British National Grid system can be seen in Figure 3.6.

By the early 1950s, large-scale (1 : 25,000) maps were available for most of Britain with the National Grid. In 1950, at a conference of the Botanical Society of the British Isles on 'Aims and Methods in the Study of the Distribution of British Plants', A. R. Clapham, then Professor of Botany at Sheffield University, gave a paper in which he suggested that the Botanical Society should embark on production of a set of distribution maps which, when produced, should be comprehensive and accurate, and that the unit of mapping should be the 10 km grid square.

The 'Maps Scheme'

Clapham's suggestions were taken up and funding obtained. In 1954 S. M. Walters was appointed Director of the project (known then as the Maps Scheme) and F. H. Perring, Senior Worker. The culmination of the scheme was the *Atlas of the British Flora* (Perring and Walters 1962).

The basis of the scheme was to indicate, by means of a map symbol, the presence or absence of a species in every 10 km grid square as defined by the British National Grid. You will recall that such a grid is reproduced on sheets of the 1 : 25,000 OS map series. There are some 3,500 such grid squares covering the British Isles. From field record cards for each grid square, and other published records, a master card was prepared and the data punched on to a single Powers-Samas forty-column card (Powers-Samas being the name of the rather bulky piece of apparatus that was adapted to do the mapping). Each species record had its own numerical code, grid reference, and vice-county code.

In the 1950s computing was rather primitive, and the data were examined in card-sorting machines that could be set to 'look for' holes punched into particular columns. The art of computer mapping was almost non-existent,

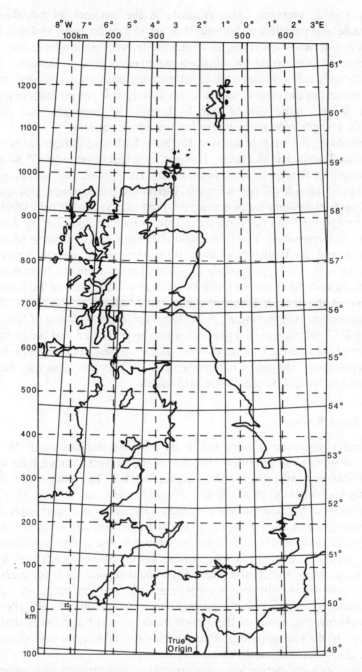

Figure 3.6 The relationship between the Transverse Mercator graticule (continuous lines) and the British National Grid System (broken lines). (Redrawn and modified from Harley 1975)

but Walters and Perring had the brilliant idea that it should be possible for the grid-reference data holes to be read by a machine and transferred mechanically to a position on a blank map. Within a few weeks, the problem had been solved by the manufacturers of the punch cards. All cards for one species were sorted on the first figure of the grid reference into four vertical strips of the British Isles, each 20 km sq wide. Within each strip the cards were then further sorted into north-to-south order. A base map was set in position in the Power Samas tabulator (rather like a mechanical typewriter) and when a species card was sensed the print-unit of the tabulator brought up a symbol into the correct position. By this means a map containing about 1,000 dots could be produced in 20 minutes. By modern standards this was very slow, but it was still a tremendous achievement for its time.

The Irish Grid

The collection of Irish data by the Botanical Society of the British Isles ran into difficulties from the very beginning of the Maps Scheme. Not only was a grid missing from half the country but there were no large-scale maps even for the gridded eastern half, the grid being an extension of the British Grid. To overcome this deficiency the British Grid was extended westward over Ireland and marked on to ½ inch : 1 ml map sheets which were provided for botanical field work. A rectangular co-ordinate system thus covered the whole of the British Isles. The main necessity for a single system lay with the rather simple method of automatic map-making whereby the apparatus could only move about in a simple rectangular co-ordinate framework – either up, down, or across the map page. Thus all the maps in the *Atlas of the British Flora* are drawn as if the British and Irish National Grids were one.

The extension of the British National Grid system to cover Ireland out-raged several Irish cartographers, and there followed a heated exchange in the scientific literature, not least by Niall MacNeill, who in the early 1960s was head of the Ordnance Survey of Ireland. Quite rightly, MacNeill pointed out that the relationship between the British National Grid and the Transverse Mercator projection becomes progressively distorted away from the central meridian. Thus the whole idea of mapping presence and absence on a fixed areal unit becomes a nonsense. MacNeill's criticisms were timely because, in 1964, the Nature Conservancy in Britain set up the Biological Records Centre at its Monks Wood Experimental Station. The Biological Records Centre was to act as a focus for national distribution mapping schemes and the lack of cartographic facilities in Ireland meant that the Centre had to produce distribution maps for the whole of the British Isles. MacNeill was determined that the matter be put right, although Perring (1968a) put up a stout defence for the single grid system, which he called the *Biological Grid*.

In fact, MacNeill (1968) in his rejoinder to Perring, pointed out that the Irish Grid had recently been published on several map series and, with a plea from the heart, urged that Irish and visiting naturalists give their map references only in terms of the Irish Grid. Presumably, MacNeill's plea did not fall on deaf ears because since the early 1970s, and the replacement of the tabulator printer by more modern graphics software and hardware, the Biological Records Centre has produced maps which show the Irish Grid correctly aligned and centred.

Grid mapping and atlases

The *Atlas of the British Flora* paved the way for a gamut of similar publications for other taxa, many of which have been published by the Biological Records Centre (Harding 1985) although some, such as *The Atlas of Breeding Birds of Britain and Ireland* (Sharrock 1976) have been published commercially. In the last few decades there has also been a growing interest in county atlases, and even urban atlases, based on the gridding methods. For this type of atlas the 10 km grid square is too large and recording is often based on a 2 km grid square called a tetrad. This mapping unit was employed for the first computer-mapped local atlas, that of the flora of Warwickshire (Cadbury, Hawkes, and Readett 1971). The derivation of the tetrad grid is shown in Figure 3.7.

The production of an atlas usually requires the labour of many amateur recorders, as well as professional scientists. The co-ordination of the publication is much better served by using grid squares than a vice-county scheme, since squares can be searched much more systematically. Scientists at the British Biological Records Centre also suggest that the grid system allows repeated sampling of a relatively small area, as a grid square is rather akin to the quadrat used by ecologists in their fieldwork. It is intended to repeat national surveys every 20–25 years at the 10 km square scale, at least for major taxonomic groups.

There are those who would rather forget that grids had ever been invented. A valid point, argued strongly by Webb (1980), is that at least from a vice-county list, one can develop a mental map of the distribution because of the way vice-counties are numbered, whereas a lifetime spent with grid references is not likely to produce the same results. As Dudley Stamp once pointed out: 'nature abhors straight lines and undoubtedly detests kilometre grids'. An example he cites of the failure of the gridded map to reveal very much biogeography is found in north-west Scotland. There the narrow outcrop of Durness limestone runs obliquely to the National Grid lines and the distribution of lime-demanding plants is only shown as an irregular pattern on a grid map (Stamp 1962).

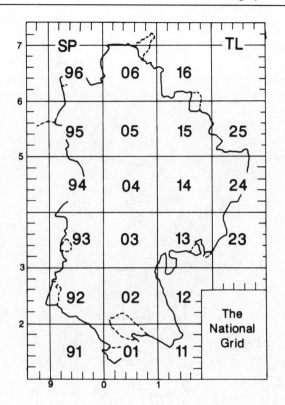

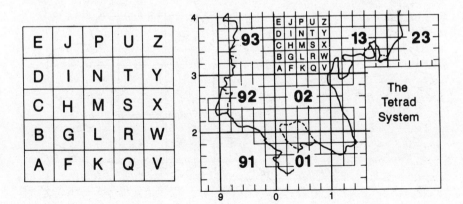

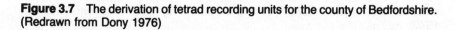

Figure 3.7 The derivation of tetrad recording units for the county of Bedfordshire. (Redrawn from Dony 1976)

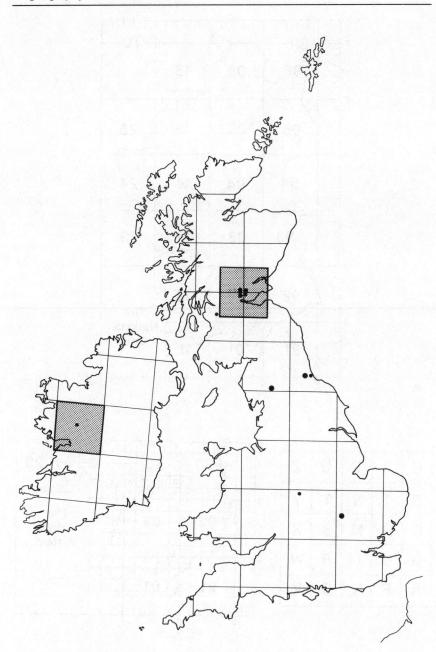

Figure 3.8 Distribution of the rare, black-necked grebe (*Podiceps nigricollis*). Within shaded areas dots have been placed centrally. (Redrawn from Sharrock 1976)

Although grid mapping is now the dominant form of biogeographical mapping in the British Isles, the dominance being partly due to the ease of computer-mapping such data, there are difficulties which should not be overlooked. In some terrain, it is just as difficult to reference a species location in terms of an easting and northing as it is to decide which of two vice-counties it is in. There are also bound to be recording errors along grid margins.

A second point to be made is that although modern grid maps are drawn by computers, this does not mean that the dots are always placed meaningfully in the grid square. By convention the dot is placed in the centre of the grid square. This does not matter too much at the scale of the British Isles, but for more local mapping, particularly in very varied terrain, one cannot deduce anything about habitat from the position of the dot. A further consideration which deserves mention is the deliberate misplacement of dots in some atlases so as not to give away the locations of rare species. In the *Atlas of the British Flora*, for example, dots of a few rare species were displaced by one 10 km square in any direction, but we are not told which species are involved. A slightly more useful approach for the mapping of rare species was adopted in *The Atlas of Breeding Birds in Britain and Ireland*. Here three devices were used: (1) dots were moved by one or two 10 km grid squares; (2) dots were placed centrally in shaded 50 km or 100 km squares; (3) dots were omitted entirely. In all three cases, however, these protective measures have been indicated alongside the map (see Figure 3.8).

A map is, of course, only as good as its raw data. In the case of grid

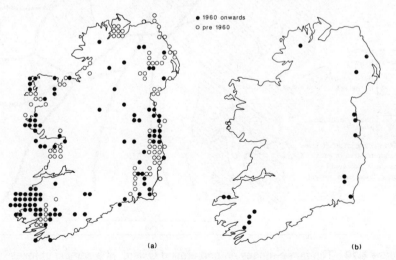

Figure 3.9 The early bumblebee (*Bombus pratorum*): (a) distribution of 10 km grid squares examined by recorders; (b) presumed distribution. (Based on Alford 1973)

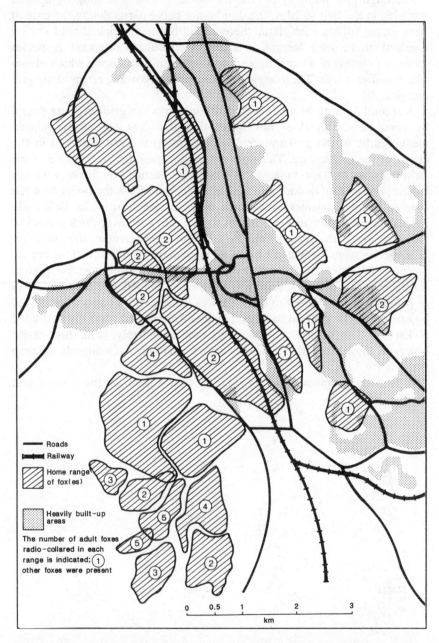

Legend within figure:

— Roads

▶▬◀ Railway

▨ Home range of fox(es)

▦ Heavily built-up areas

The number of adult foxes radio-collared in each range is indicated; ① other foxes were present

0   0.5   1   2   3
km

Figure 3.10 The home-ranges, in and around Oxford, of a sample of foxes (*Vulpes vulpes*) fitted with radio-collars. Numbers indicate the number of radio-collared foxes in each range. (Redrawn and modified from MacDonald and Newdick 1982)

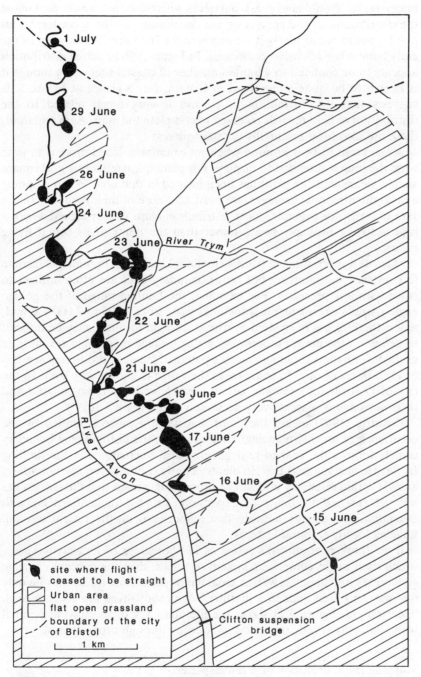

Figure 3.11 The linear range of the small white butterfly (*Pieris rapae*) in Bristol. (Modified and redrawn from Baker 1982)

mapping, we should always ask ourselves whether or not we are looking at the distribution of the species or the distribution of the recorders? This point is made rather nicely if we examine the Irish distribution map of the early bumblebee (*Bombus pratorum*). In Figure 3.9b its actual distribution appears to be confined to a limited number of coastal sites, even though it is known to be quite widespread in Britain. But a glance at Figure 3.9a suggests that its distribution in Ireland is very much related to the distribution of field records received to complete the map. As is apparent, these are also from essentially coastal squares!

A last point to bear in mind when we examine gridded maps is to note what the map is intended to display. For plants, a symbol in a grid square simply means that the plant has been located in that unit. When mapping animal distributions, however, we must take note of their mobility. In the case of birds, for example, most distribution maps usually record where a bird nests, or probably nests, rather than the limits of its actual flying range.

Maps of a species' range (its distributional area) are, in principle, simple to construct and interpret but range boundary lines are often difficult to place with any precision. For example, the hunting range of the golden eagle (*Aquila chrysaetos*) is sometimes vast, from 5,000 to 7,000 ha, and, unless the bird has been sited overhead its flight location, may be difficult to define on the map. Futhermore, a single-range map might reveal little useful information about an animal's potential location at different seasons or even times of day. In some cases the range is known very well. In the city of Oxford, for example, urban foxes have been radio-collared and their movements mapped very accurately (see Figure 3.10).

Udvardy (1969) makes the point that when a range is limited by the occurrence of an environmental limiting factor, the distributional boundary usually follows the year-to-year geographical fluctuations of the limiting factor. He also suggests that, in plants, a fairly continuous range boundary, at least within relatively uniform environments, means that the species has not yet reached its limits of expansion. Fringed range borders, with discontinuities, usually indicate that a species is on the retreat but still utilizing locally favourable areas.

If we think of a animal's range as the boundary of its activity space, which itself might comprise a system of tracks, then the concept of range generalizes into linear as well as areal types, since some tracks may be only one-way and used once in a lifetime. Many short-lived insects have such linear ranges. Figure 3.11 shows the lifetime track of an individual, small white butterfly (*Pieris rapae*) on its journey through urban Bristol.

Mapping the distribution of marine species

The mapping of marine species around the British Isles has a long and distinguished history but is not as well known or as fully developed as terrestrial mapping schemes. There are two basic reasons for this. First, apart from bottom-dwelling organisms, marine taxa are located in a three-dimensional space, which is for the most part out of view, although some data are now collected by divers. Second, casual observations are rare and sometimes have no precise location.

In the nineteenth century some general ideas of British marine bio-geography had emerged. This was mainly through the activities of the so-called Dredging Committees whose members hired boats and scooped up what they could for identification. Later, this was achieved through the establishment of a number of Marine Biological Stations. Since the foundation of the International Council for the Exploration of the Sea (ICES) in 1903, commercial fish stocks in the seas around the British Isles have been closely monitored, particularly by the Directorate of Fisheries Research at Lowestoft, which is a laboratory of the Ministry of Agriculture, Fisheries, and Food. In 1981 the Directorate published an *Atlas of the Seas Around the British Isles* which has very useful maps of the distribution of economically important fish species (Lee and Ramster 1981). A new edition of this atlas is currently being prepared and a version will also be available on floppy disc for viewing on a microcomputer.

The British marine census areas

The Conchological Society of Great Britain and Ireland was probably the first learned society to propose marine mapping units, and as early as 1901 had defined the *British Marine Area* on which distribution maps of marine molluscs were later based. The definition of the mapping units was in places inadequate and as a consequence the limits were redefined in 1961 resulting in the *British Marine Area 1961*. This comprises the British Marine Area within the 100 fathom isobath together with the North Atlantic Area beyond 100 fathoms (see Figure 3.12). There are forty-two census units altogether, and the distribution maps produced from these data show the presence of a live post-1950 records by a solid dot placed more or less centrally within the census unit.

The census units themselves vary considerably in size, and their bound-aries are defined mostly by meridians and parallels on what appears to be a Mercator projection. Although this type of dot map is based on an irregular grid, such maps can give the Conchological Society a reasonable impression of mollusc distribution. On the other hand a dot in census area 24 (Liverpool Bay) clearly reveals more spatial information than one in, say, area 4 (Viking) which is about fifteen times as large. Notwithstanding this reservation and the difficulties of obtaining good data, very interesting distribution maps have been constructed, notably in the *Sea Area Atlas of the Marine Molluscs of Britain and Ireland* (Seaward 1982).

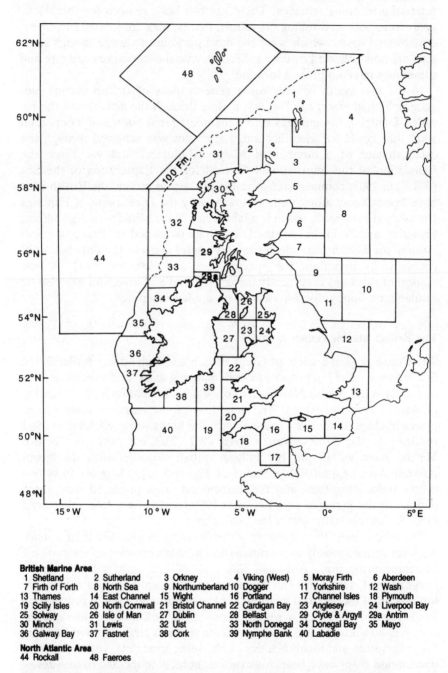

Figure 3.12 The British Marine Area, 1961. (Based on Seaward 1982)

British Marine Area

1 Shetland	2 Sutherland	3 Orkney	4 Viking (West)	5 Moray Firth	6 Aberdeen
7 Firth of Forth	8 North Sea	9 Northumberland	10 Dogger	11 Yorkshire	12 Wash
13 Thames	14 East Channel	15 Wight	16 Portland	17 Channel Isles	18 Plymouth
19 Scilly Isles	20 North Cornwall	21 Bristol Channel	22 Cardigan Bay	23 Anglesey	24 Liverpool Bay
25 Solway	26 Isle of Man	27 Dublin	28 Belfast	29 Clyde & Argyll	29a Antrim
30 Minch	31 Lewis	32 Uist	33 North Donegal	34 Donegal Bay	35 Mayo
36 Galway Bay	37 Fastnet	38 Cork	39 Nymphe Bank	40 Labadie	

North Atlantic Area

44 Rockall	48 Faeroes

Continuous plankton recorder data

The continuous plankton recorder, designed by Hardy in 1936, is used behind commercial freighters on their normal sailing schedules. It is towed at a depth of 10 m, sampling the plankton continuously on a moving band of silk mesh. From 1932 to 1939 it was used mostly in the North Sea, and after the Second World War the sampling was gradually extended to include a comprehensive cover of the north-eastern Atlantic. This was made possible by the introduction of the ocean weather ships which began to tow the recorders in 1948. The organization of the data was originally carried out by the Department of Oceanography in the University College of Hull, as it then was. In 1950 the responsibility passed to the Oceanographic Laboratory in Edinburgh, run by the Scottish Marine Biological Station. The whole collecting scheme is now known as the Continuous Plankton Recorder Survey (Colebrook *et al.* 1961). As far as possible, every recorder route was sampled once a month. The records were divided into sections, each representing about 15 km of towing, and each assigned to a geographical unit based on a 1° longitude by 1° latitude rectangle on, probably, a Mercator projection. Whatever the projection might be, it is clear from an inspection of the maps produced from the recorder surveys, that the mapping unit becomes progressively larger in a northerly direction and is not an equal-area projection. However, the distortion from a true grid is only slight, and because data were collected regularly over a long period not only was it possible to map presence and absence but also to map levels of abundance. Because a straight line is a line of constant bearing on the Mercator projection, the latter is popular for navigation charts.

The organizers of the survey attempted some very ambitious mapping of their grid-based data. Small dots were used to indicate the centre of a sampled rectangle. Large dots were used to show that a particular species of plankton was present in the sample. The numerical abundance of each species in the sample was calculated and the data contoured using a logarithmic interval. In Figure 3.13 two isopleth maps from the Survey are reproduced and clearly show the value of such mapping. The map for the diatom *Thalassiothrix longissima* shows the remarkable correspondence of the contours of abundance and the position of the continental shelf, *Thalassiothrix* being abundant in deeper water but scarce to absent in less saline, shallower water. On the other hand, the zooplankton *Centropages hamatus* is most abundant in the southern North Sea, away from the tongue of saline Atlantic water which extends into the northern North Sea basin.

The UTM grid

Neither the British marine census area maps nor those used for recording the Continuous Plankton Survey data, are really satisfactory from a

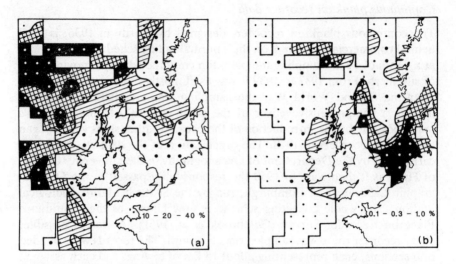

Figure 3.13 Species-abundance maps redrawn from the Continuous Plankton Recorder Survey: a) *Thalassiothrix longissima*; b) *Centropages hamatus*. All sampled squares are indicated by a dot. Those squares containing the plankton have then been contoured and density shaded according to the percentage levels shown in diagrams. (Redrawn and modified from Colebrook *et al*. 1961)

cartographic point of view because of the distortion due to the map projection. To overcome this problem, a number of biogeographical atlases have adopted the Universal Transverse Mercator projection (UTM) for map construction. The UTM grid is defined on the projection by using the equator as the northings origin, and the central meridian of the zone in question (which might be the British Isles) as the eastings origin. A metric grid is then constructed both parallel, and at right angles, to the central meridian. This grid system provides recording units of equal area throughout its latitudinal range.

The size of the UTM grid is arbitrary and depends on the mapping project. Each grid square can be uniquely identified by a letter code. The use of the UTM system is illustrated in Figure 3.14 which shows the westerly distribution of the dinoflagellate *Ceratium compressum* plotted as open circles positioned in the centre of 100 km grid squares. The actual grid lines are not shown in this diagram. The system is described further in Dodge (1981).

The UTM grid has been adopted for many European mapping schemes, such as the European Plant Parasitic Nematode Survey (Alphey and Taylor 1986) and for the maps in the Atlas Florae Europaeae project. One such map from this project is shown in Figure 3.15. On this map dot symbols representing the presence of the sea pearlwort (*Sagina maritima*) are placed in 50 km square grid squares, which is the unit usually chosen for distribution mapping at this scale.

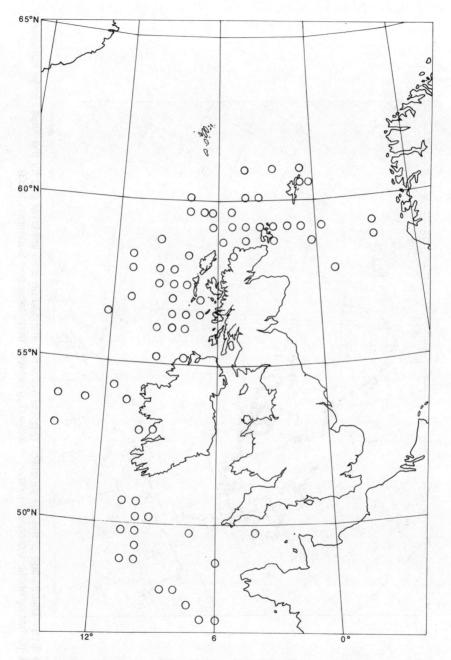

Figure 3.14 Distribution of the dinoflagellate *Ceratium compressum* recorded in 100 km grid squares of the Universal Transverse Mercator projection. (Redrawn from Dodge 1981)

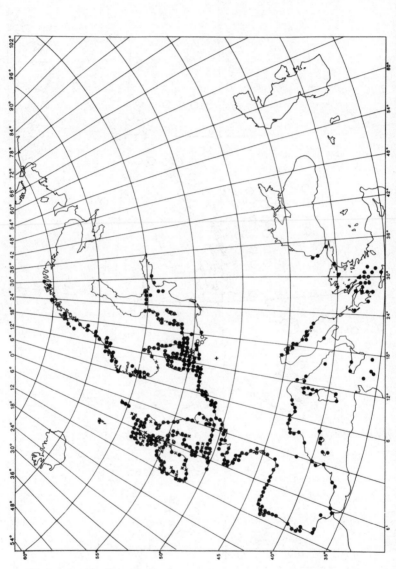

Figure 3.15 Distribution of the sea pearlwort (*Sagina maritima*) mapped on a 50 km UTM grid for the Atlas Florae Europaeae project. (Other symbols: x probably extinct; + extinct). (Redrawn from Jalas and Suominen 1983)

Problems in marine mapping

The mapping of many marine organisms illustrates well a number of more general points that we should bear in mind. Many species are very short lived, and whether or not they are recorded at a particular site depends very much on how frequently the site is visited. For example, the marine alga *Prasiola stipitata* often forms a distinctive band at the top of the shore, and is recorded around much of the coast of Britain. A distribution map for the species would show it as common on the east coast of Scotland and in the Hebrides, but missing from much of the north-western coast of Scotland. Similarly, it is found on the Devon and Cornwall coasts but not on the coasts of south-west Ireland. There do not appear to be any geographical reasons for its absence, but its ephemeral nature suggests that this alga is most probably under-recorded, and is in fact much more widespread.

Some species are difficult to map, simply because they are hidden and may not be noticed. The alga *Tellamia intricata* occurs in scattered coastal sites throughout the British Isles and is almost certainly quite common. Its under-recording is due to the fact that it is found beneath the superficial membrane covering the shells of flat periwinkles of the genus *Littorina*, and is thus difficult to detect.

One of the perennial problems in studying many marine organisms, and many small non-marine species of plants and animals for that matter, is that they are often very difficult to identify correctly. For example, it is sometimes difficult to distinguish between the species of green algae. The thin, crinkled, slimy sheets of vivid green colour often seen at low tide on rocky shore are species of sea lettuce (*Ulva spp.*). It was once thought that *Ulva rigida* was rare, and confined to southern locations in the British Isles. In fact its distribution is more widespread, and it is often overlooked because it can only be distinguished from *Ulva lactuca* under the microscope.

Chapter four

Useful concepts

So far we have seen that the dominant tradition in British biogeography has been the examination of the spatial distribution of plant and animal species. In order to make sensible interpretations of these biogeographical data it is necessary to be conversant with relevant ecological and biological concepts.

Some concepts, such as ecosystems, are dealt with at great length in many biogeography textbooks, but as our focus is on species distributions – rather than the study of the mechanisms of energy and nutrient flows within ecosystems – only essential material is described here.

Species and speciation

The basic unit of study in biogeography is the *species* which can broadly be defined as a group of morphologically (phenotypically) similar organisms which are capable of breeding with each other to produced fertile progeny, and which do not interbreed with members of other species. The term species belongs to a hierarchical system of naming organisms which owes its development to the eighteenth-century Swedish naturalist, Carolus Linnaeus. He used Latin binomials which gave plants and animals two names. The first name he termed the generic and the second the specific. Thus the sweet violet belongs to the genus *Viola* and has the specific name (or epithet) *odorata*. It is usual to capitalize the generic name, and the full binomial description might also acknowledge the scientist who first described the species e.g. *Viola odorata* L., where the L. indicates that it was first described and named by Linnaeus. Those species that share features in common are placed in a larger group called a genus (plural – genera) and similar genera can be grouped into families.

Not all species multiply by sexual reproduction and the simple definition of species given above is not totally satisfactory. For example, many primitive plants and animals reproduce asexually, and the term *apomixis* (adj. – apomictic) is given to this type of reproductive system.

Individuals in a natural population do not all have the same chance of breeding with each other. Often a population is fragmented into local breeding units called *demes*. Should the demes become isolated their members may, through natural selection, lose the ability to breed with individuals belonging to other demes. In terms of our definition, they may be thought of as a new species. The simplest type of isolation comes about when demes become geographically separated, dispersal is reduced, and the exchange of genetic material limited. The isolating factor may be a rise in sea level, a glacial advance or local tectonic movements. In the fullness of time, the isolated demes genetically differentiate to form new species. This model of speciation is called *allopatric speciation*.

A particular type of allopatric speciation, called the *founder principle*, is initiated when a few individuals form a geographically isolated population – for example, by colonizing a remote island. The colonizing individuals are likely to have only a small, and perhaps biased, sample of all the genetic variation in the parent population and thus the colony's population may develop very differently. Berry (1977) has identified the founder principle in operation in his studies of mice on islands around the British coast. The house mouse, *Mus musculus*, was introduced to the island of Skokholm, off the Welsh coast, in stores taken to the island by farmers at the turn of the century. Berry used measures of genetic similarity to demonstrate that Skokholm's population of mice is different from populations on the nearby Pembrokeshire mainland.

A second type of speciation, known as *sympatric speciation*, occurs within spatially continuous populations. Two types of mechanism may bring this about. The first occurs where there are strong selective pressures causing the population to adapt to local environments. An alternative mechanism is by chance chromosomal changes, especially through the doubling or tripling of their number, a process called *polyploidy* which is fairly common in plants. Polyploidy is significant for the biogeographer since such species are often vigorous and competitive. A well-known example of speciation through polyploidy is seen in the grass *Spartina*. In 1870 a new species of *Spartina* (*Spartina townsendii*) was observed in Southampton Water. This is a sterile hybrid cross between the American species *Spartina alternifolia* and a native species *Spartina maritima*. By 1890, a vigorous fertile hybrid appeared, called *Spartina anglica*. It is thought that this hybrid was derived from *Spartina townsendii* by a doubling of the number of chromosomes. (Marchant 1967).

Habitat

The term habitat literally means the place where an organism or community lives, or where one might expect to find it. As such, it can be described in terms of measurable environmental characteristics, such as exposure, soil type, and water temperature. For example, the habitat of the

75

salt marsh community is that part of the coastal strip which is periodically inundated by salt water. The idea of the habitat as an 'address' of an organism is a useful one. But if we want to know about an organism's activities and its relationships with other organisms then we must also know something about the organism's *niche*, or as the American ecologist E. P. Odum would say, 'its profession'.

Few attempts have been made to classify habitats. One reasonably comprehensive classification is that of Elton and Miller (1954), who divided the biosphere into seven *habitat systems*. Each habitat system may then be further divided into *formation types* and stratified into *vertical layers* with a small series of *qualifiers* to take care of some other important division of the systems and formation types.

(i) *subterranean system* – includes all those habitats below the soil and subsoil such as underground caves. All species are dependent on organic waste products washed in from above or brought in by bats or birds.

(ii) *high air system* – no animal or plant spends its whole life in this system but it may be used during migration and dispersal.

(iii) *domestic system* – includes those habitats influenced by man's activities, and includes sites such as rubbish tips, sewers, buildings, and gardens.

(iv) *general system* – consists of small, decaying organic habitats scattered throughout the other habitat systems and includes animal artifacts such as nests and burrows, human artifacts such as haystacks and silos, dead and dying wood, dung, and carrion.

(v) *aquatic system* – includes all water bodies, whether natural or artificial, their margins being usually separated from the dry land by a transition belt. Elton and Miller recognize twenty-five formation types based on size of water body and speed of water movement (see Table 4.1) although not all occur in the British Isles.

(vi) *transition system* – occupies the most striking of all surface boundaries, the zone between land and water. This system comprises many types of habitat usually classified according to the nature of the adjacent water body. They include marshes, the edges of lakes, ponds, rivers, streams, salt marshes, and the intertidal seashore zone.

(vii) *terrestrial system* – ranges vertically from the subsoil up to the top of the high tree canopy. Four formation types are recognized (woodland, scrub, field, and open ground) and these correspond more or less to an ecosystem succession from bare ground to climax woodland. For the open ground and field formations, four qualifiers can apply. These are: acid, non-acid, maritime, and arable. For the woodland and scrub formations there are three possible qualifiers: deciduous, conifer, and mixed (deciduous and conifer). A final and most

Table 4.1 Formation types of the aquatic habitat system

Speed of flow	Size of water body				
	A (very small)	B (small)	C (medium)	D (large)	E (very large)
1. still	tree-hole	small pond < 17 m²	pond < 0.4 ha	large pool or tarn < 40 ha	lake or sea
2. slow	trickle shallow stream	ditch, field dyke	canal, river back-water		
3. medium	trickle shallow stream	lowland brook or small stream	lowland river	lowland large river	river estuary
4. fast	spring	upland small torrent, stream	large torrent, stream		
5. vertical or steep	water drip cascade	small weir or waterfall	large weir or medium waterfall	large waterfall	—

Source: Elton and Miller 1954

important qualifier is the *edge*. This is used mainly for woodland and scrub formations, and is a narrow transition zone allocated to the higher formation type. It is usually more species rich than the heart of a closed scrub or woodland. The terrestrial formation types, their vertical layers, and edges are shown in Figure 4.1.

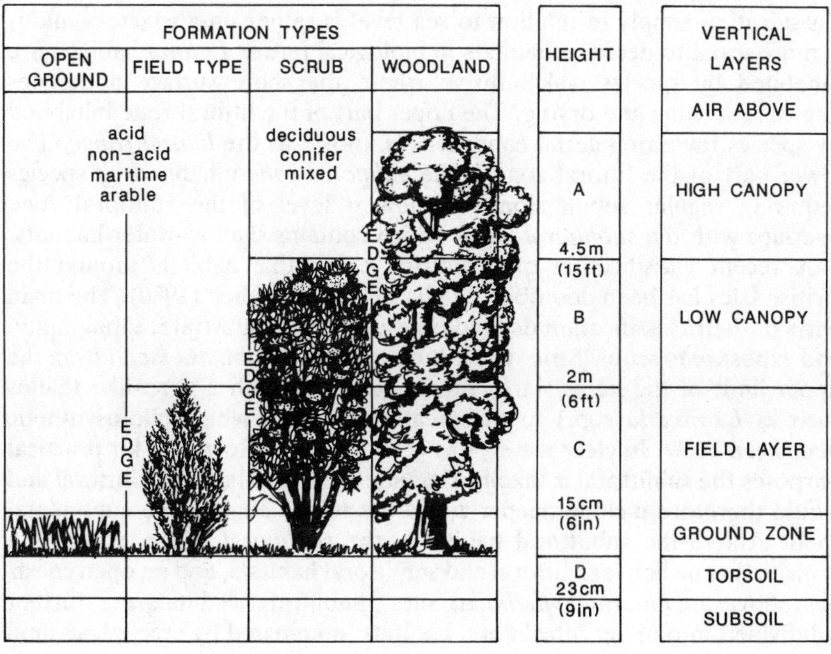

Figure 4.1 Terrestrial habitat classification. (Modified from Elton and Miller 1954)

Elton and Miller intended that their classification be practical and a simple coding system was devised. The vertical layers of the terrestrial system are first assigned letters (see Figure 4.1). Subscripts are then used to denote the number of under-layers present in any vertical section. For example, A_2 is the code for the upper storey, or 'high canopy', of a woodland having two under-layers. The whole vertical profile might be coded as: $A_2{}^{30} A_1{}^{20} C^1$, which would describe a woodland habitat having an upper storey of 30 m, a lower storey of 20 m, no lower canopy, a 1 m high field layer, and no plants in the ground zone.

Marine habitats are not well covered by Elton's scheme, and more usually they are sub-divided into two basic types, *pelagic* and *benthic* (see also Chapter two). Pelagic habitats are not usually further classified since there is no sense in which we can talk about the habitat as a place in the context of open water. Many organisms in the sea drift with the currents (the plankton) and free-swimming organisms (the nekton) often range over large areas. The pelagic zone of the seas around the British Isles is less than 200 m deep, and at the global scale is described as epipelagic to indicate it is really only the upper layer of the pelagic zone proper. Benthic habitats are sometimes classified according to the degree of exposure to the air. *Supratidal* habitats are never completely submerged by the tides, while those which are never exposed to the air are known as *subtidal* habitats. Intermediate between these two habitats is the *intertidal* habitat. A classification simply in relation to sea level is rather unsatisfactory and it is more usual to describe habitats in biological terms. *Littoral* habitats are inhabited by species which thrive where the shore surface undergoes alternate wetting and drying. The upper part of the littoral zone inhabited by species favouring aerial conditions is known as the *littoral fringe*. The lower part of the littoral zone is the *eulittoral zone* inhabited by species requiring regular submersion. The lowest level of the eulittoral zone overlaps with the *sublittoral fringe* which contains shallow-water habitats.

A recent classification of sublittoral epibenthic habitats around the British Isles has been described by Hiscock and Mitchell (1980). The main parts of their classification depend on water depth, substrate, topography, and exposure to scour. Strictly speaking *sublittoral habitats* extend from the upper limit of the kelps (large brown seaweeds with a strap-like thallus such as *Laminaria* spp.) to the greatest depth at which photosynthetic plants can grow. In clear water, this is about 30 m. However, for practical purposes the sublittoral is taken to include all depths below the littoral and would therefore include deeper zones out to the edge of the continental shelf. Above the sublittoral habitat is the *sublittoral fringe*, which is a transition zone between littoral and sublittoral habitats, and on open coasts lies above *Laminaria hyperborea* sites. Sublittoral habitats are further subdivided into upper *infralittoral* habitats, dominated by erect algae, and lower *circalittoral* habitats dominated by animals rather than photosynthetic algae. Each major depth zone is then further categorized according to the substrate (bedrock, boulders, or pebbles) and topography (cliffs, over-hangs, gullies, etc.). For practical purposes, the description of marine habitats has to be kept simple because of the problems of recording data in these environments.

A different approach to the classification of habitats was described by Southwood (1977), who suggested that they might be classified not in the way we ourselves view the habitat, but from the point of view of the organism itself, both in terms of space and time. For example, the habitat of a browsing deer, with a spatial range encompassing many types of shrubs and trees on which it feeds indiscriminately, is relatively stable and predictable. The 'same' habitat for a small insect, dependent on just one type of plant fruit, might prove totally unstable and unpredictable if its one supply of fruit is destroyed. Southwood's ideas are illustrated in a simplified form in Figure 4.2. The alphabetical notation in this diagram is defined as follows:

r   favourableness, the length of time permitting breeding; the intrinsic rate of increase that an individual might expect to attain in that habitat.

H   length of time that a location permits existence.

L   length of the unfavourable period

T   the generation time of the organism

R_t the trivial, day-to-day, feeding range of the organism

R_m the migrating range of the organism

U   size of unfavourable area between habitats

In Figure 4.2, the upper diagram shows the spatial characteristics of three types of habitat. In continuous habitats, r values vary little as compared with R_t. At the other extreme, isolated habitats have U values that generally exceed even the migratory range of the organism, R_m. In the lower diagram, four types of habitat are defined in terms of their time characteristics. Habitat favourableness, r, may vary little, as in constant environments, or periodically, in which case the habitat has a seasonal environment. In unpredictable environments, the r value may vary con-siderably, as does L, though they are both short enough, as compared to T, for H to be at least moderate. In ephemeral environments, the length of the unfavourable period, L, is predictably long and H, therefore, is short. Over time, of course, life histories become suited to habitats through the process of natural selection and organisms themselves may influence the nature of their habitat. One important aspect of Southwood's approach to habitat classification is that it provides us with a framework within which we can think about, for example, land-use changes (man-made changes in U) and climatic changes (possible changes in L and H), and their impact on the distribution of organisms. The match between life history, strategy, and habitat characteristics of an invading or dispersing plant or animal very much determines its successful establishment.

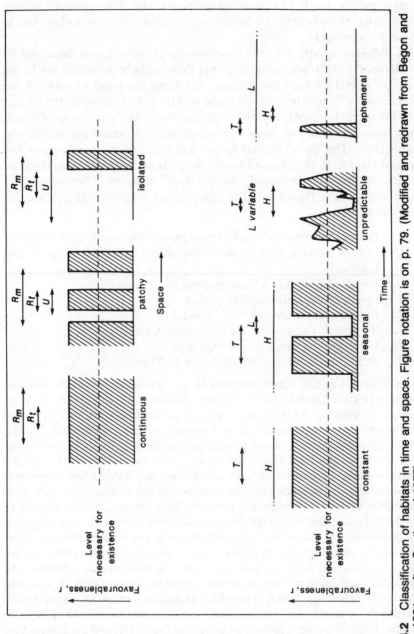

Figure 4.2 Classification of habitats in time and space. Figure notation is on p. 79. (Modified and redrawn from Begon and Mortimer 1981 – after Southwood 1977)

Resource allocation models

In order to complete their life cycles successfully, organisms have evolved general budgeting mechanisms for the utilization of energy and available resources. For example, in plants, the amount of photosynthetic energy allocated to roots, leaves, and reproductive organs, and the amount of time spent in dormancy, growth, and maintenance, are important attributes that govern success. Different resource-allocation strategies are applicable in different habitats, and form a major part in the competitive performance of a species. There is a spectrum of strategies, two ends of which are called *r selection* and *K selection* (Table 4.2). The letters r and capital K refer to parameters in the logistic equation for population growth; r controls the rate of population increase, in other words how quickly the initial part of the growth curve climbs, and K is a ceiling on the population and can be thought of as the level at which the curve flattens out. K is often thought of as the carrying capacity of a habitat, since K-selected species live in habitats which exhibit density dependent mortality.

Table 4.2 Characteristics of r- and K-selected species

r species	K species
Small and short lived	Large and long lived
High fecundity	Low fecundity
Short generation time	Long generation time
High rate of dispersal	Low rate of dispersal
Very variable population density	Stable population density
Time efficient	Food and space resource efficient
Opportunistic, exploiting temporary habitats in variable climates	Equilibrium species of stable habitats in predictable climates
Productive	Efficient

Source: Southwood (1977)

In plants, K-selected species tend to be long-lived, have a prolonged vegetative stage, allocate a small fraction of energy to reproduction and tend to occupy late stages of ecological successions. At the other extreme, r-selected plants are short lived and allocate a high proportion of available resources to reproduction and occupy early stages of successions. Such species are often described as opportunistic or colonizing (Gadgil and Solbrig 1972). The biogeographical implications of r- and K-selection are considerable. For example, in climates which vary seasonally and survivors recolonize habitats each spring, species which are r-selected are favoured. In more uniform, predictable climates, the greater efficiency of K-selected species means that they should be more competitive (MacArthur and Wilson 1967).

More recently, Grime (1979) suggested that resource allocation strategies in plants were controlled by two basic factors: stress and disturbance.

Stress is defined as any external factor that limits production, such as low levels of nutrients or light. Disturbance is defined as the partial or total destruction of the plant biomass: for example, by fire or grazing. If we only consider high and low values of stress and disturbance, the combination of Grime's factors gives rise to the following strategies:

	Intensity of stress	
Intensity of *disturbance*	high	low
high	mortality	ruderals
low	stress tolerators	competitors

Grime suggests that ruderal and stress-tolerant strategies correspond, respectively, to the extreme of r- and K-selection and that competitors occupy an intermediate position. Stress-tolerant species reduce allocations toward vegetative growth and reproduction, and exhibit features that ensure the endurance of mature individuals in harsh environments. Stress may be low temperature (as in the case of arctic-alpine habitats) or light (as in the case of shaded woodlands). In woods, species often produce a paucity of flowers and seeds unless they are in an edge habitat: for example, ivy (*Hedera helix*) and honeysuckle (*Lonicera periclymenum*). Competitors maximize the capture of resources in productive but relatively undisturbed environments. Typical competitors are rose-bay willow-herb (*Chamaenerion angustifolium*) and bracken (*Pteridium aquilinum*). Both plants have large reserves of energy stored in underground organs which can be mobilized readily during the growing season to support the rapid expansion of foliage and thus utilize space in the canopy environment. Ruderals are usually herbs having a short life-span and high seed production. They are often found in highly disturbed, but potentially productive, environments such as the drift line on the shore, where species such as the saltwort (*Salsola kali*) and pineapple weed (*Matricaria matricarioides*) may be rooted in loose organic debris. Arable farmland and trampled ground are other typical environments where ruderal species flourish. Grime's classification has also been applied to marine environments by Dring (1982). For example, fugitive species of algae, such as those belonging to the genera *Ulva* and *Enteromorpha*, which rapidly colonize disturbed sites, are equivalent to Grime's ruderal type. Species of *Pelvetia* and *Ascophyllum* can withstand desiccation and low light levels, and can be thought of as stress tolerators. The great majority of *Laminaria* species have broad strap-like thalli which rapidly form an effective canopy which shades out understorey species. They are thus classed as competitors.

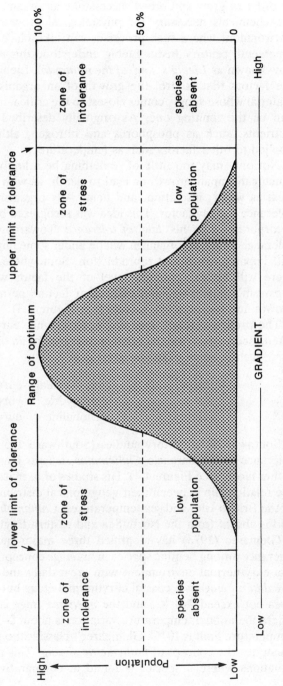

Figure 4.3 Limits of tolerance along a gradient of a physical factor.

Tolerance

In order to grow and breed successfully an organism must have access to the chemicals necessary for physiological processes and also be in an environment where such processes can take place. In the middle of the nineteenth century Justus Liebig understood this and formulated what is now known as *Liebig's 'law' of the minimum*. Liebig's idea was that of all the factors that control the growth of an organism, it is that essential material whose supply comes closest to the critical minimum needed which will be the limiting one. As originally described, the 'law' applied to nutrients, such as phosphorus and nitrogen, although it is sometimes applied to other factors such as temperature.

Not only may too little of something be a limiting factor but also too much can impair growth or lead to death, as with environmental factors such as water, insolation, and heat. Any organism thus has a range of tolerance for any factor. This idea was recognized by Shelford in 1913 and incorporated into his *'law' of tolerance*. Towards the upper- and lower-tolerance limits, an organism will be under some physiological stress which will impair growth and reproduction. Somewhere between these limits, there will be an optimum level of the factor which is suited to the organism's physiology and will, other factors permitting, allow maximal growth and reproductive success (see Figure 4.3).

The prefixes 'steno', meaning narrow, and 'eury', meaning wide, are often used to describe the relative tolerance of an organism with regard to some factor. Some examples are:

temperature	–	stenothermal	eurythermal
water	–	stenohydric	euryhydric
salinity	–	stenohaline	euryhaline

Southward's laboratory studies (Southward 1958) on the temperature tolerance of the barnacle *Chthamalus stellatus* illustrates Shelford's 'law' rather nicely (see Figure 4.4). His studies of barnacles clearly demonstrate the relationship between their geographical distribution around the coasts of the British Isles and sea temperatures. *Chthamalus* is a southern species and is absent from the North Sea and eastern English Channel.

Glemarec (1973) has identified three major patterns of temperature tolerance among benthic species in his *étage* concept. The *infralittoral étage* is a eurythermal environment with great daily and seasonal changes, the *circalittoral étage* is a coastal eurythermal étage but the seasonal variation does not exceed 7–8 °C, and the *open sea étage* is a stenothermal étage where the annual temperature variation is about 2–3 °C, and whose upper temperature limit is 10 °C. Glenmarec's classification is interesting in that it illustrates the concept of *bathymetric sliding*. This phenomenon describes situations where a species is found at comparatively shallow depths in

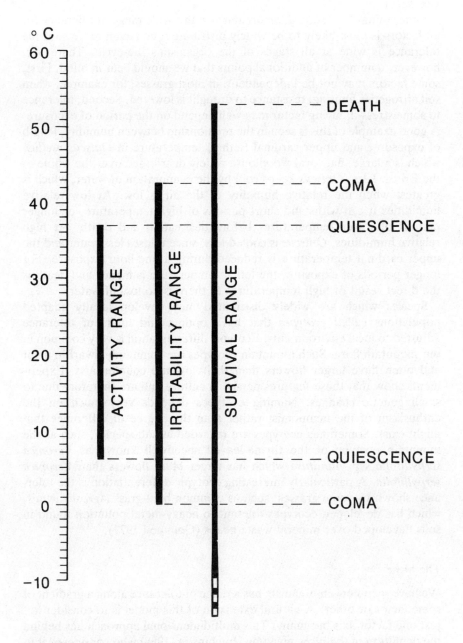

Figure 4.4 Three ranges of temperature tolerance of the barnacle *Chthamalus stellatus*. The lower death limit has not been determined. (Redrawn from Southward 1958)

northern latitudes, and only much deeper in more southerly latitudes and vice versa.

Other things being equal, an organism with a wide range of tolerance for all factors is most likely to be widely distributed, and even more so if the tolerance is wide at all stages of the organism's life-cycle. There are, however, a number of additional points that we should bear in mind. First, some factors may not be independent: in most grasses, for example, when soil nitrogen is limiting, resistance to drought is lowered. Second, tolerance to some stress-inducing factor may well depend on the period of exposure. A good example of this is seen in the relationship between humidity, length of exposure, and upper cardinal (lethal) temperature in *Oniscus asellus*, which is a large, flat, oval woodlouse widely distributed over the whole of the British Isles. *Oniscus* keeps cool by the evaporation of water, which is greatest when the relative humidity of the air is low. At low relative humidities it can withstand short periods of high temperature, or longer periods of lower temperatures before desiccation and death. At high relative humidities, *Oniscus* is cooled less, since it loses less water and the upper cardinal temperature is reduced during a one-hour exposure. For longer periods of exposure, the lethal temperature is higher and death is the direct result of high temperatures as there is no loss of water.

Species which are widely distributed may develop locally adapted populations called *ecotypes* that have optima and limits of tolerance adjusted to local environments. Ecotypic differentiation is very common in our mountain flora. Such mountain ecotypes are commonly dwarf in habit and often have larger flowers than their lowland counterparts. Experiments show that these features persist in culture and are therefore due to small genetic changes. Naming ecotypes depends very much on the enthusiasm of the taxonomist rather than the degree of difference that might exist. Sometimes ecotypes are regarded as sub-species, such as the mountain ecotype of the thyme-leaved speedwell known as *Veronica serpyllifolia* ssp. *humifusa* which has larger bluer flowers than *Veronica serpyllifolia*. A particularly interesting ecotypic differentiation is the tolerance shown by some grasses, such as common bent-grass *Agrostis tenuis*, which has developed ecotypes tolerant to heavy-metal pollution found in soils developed over mineral waste heaps (Gemmell 1977).

The niche

We have seen how an organism has a range of tolerance along a gradient of some factor or other. A natural extension of this model is to consider not just one factor axis but many. This multidimensional approach lies behind the definition of the *niche* given by Hutchinson (1958) who considered it to be defined by the total range of environmental variables to which the species must be adapted, and under which the species population lives and

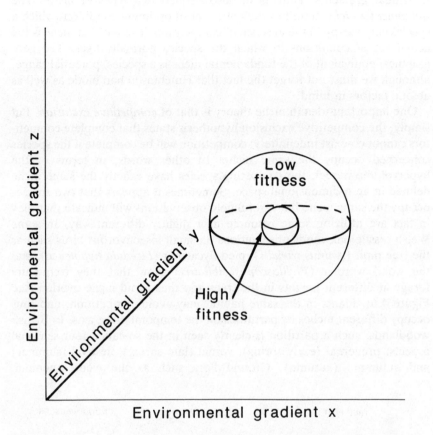

Environmental gradient x

Figure 4.5 A niche shown as a three-dimensional hypervolume. Parts of the hypervolume with high fitness (or performance) are relatively optimal for the organism concerned. (Modified from Pianka 1983)

reproduces itself. One of the nicest descriptions of the niche is by the English zoologist Charles Elton, who said that when a naturalist exclaims:

'there goes the badger', he should have in mind a picture of its function and status in the animal community, comparable to his knowledge of human society when he says, 'there goes the vicar'.

At low dimensions, it is possible to portray Hutchinson's ideas diagrammatically. In Figure 4.5, three environmental axes are drawn at right angles to each other. Within the three-dimensional space we can locate a species' tolerance limits in a continuous geometrical shape which we can call a hypervolume. (In Figure 4.5 it happens to be a sphere for the sake of simplicity.) More compact shapes are the result of peaked curves along the individual gradients. Hutchinson also defined two types of niche. The *fundamental niche* describes the entire set of optimum conditions which a species can occupy in the absence of competition. The *realized niche* is the actual set of conditions in which the species normally lives. The geographical equivalent of the fundamental niche is a species' potential range, although we must not forget the fact that Hutchinson had biotic as well as abiotic factors in mind.

One important idea in niche theory is that of *competitive exclusion*. Put simply, the competitive exclusion hypothesis states that complete competitors cannot co-exist indefinitely; competition will be complete if the species concerned occupy the same niche. In other words, in terms of the hypervolume model, the competing species have exactly the same niche defined in an n-dimensional space. Sometimes it appears that two species occupy the same niche but careful field observations will indicate that they in fact are utilizing some resource in a slightly different way. In some Welsh sessile oakwoods observations on small insectivorous birds such as the tree pipit (*Anthus trivialis*), pied flycatcher (*Ficedula hypoleuca*), and the wood warbler (*Phylloscopus sibilatrix*) show that they frequently forage at different heights in the trees and thus avoid niche overlap (see Figure 4.6). Plants in the same habitat may avoid competition, and thus occupy different niches by partitions on the temporal niche axis. In British woodlands, such a partition is clearly seen in the so-called four seasonal aspects: prevernal (early spring); vernal (late spring), aestival (summer) and autumnal (autumn). Ground flora such as the wood anemone

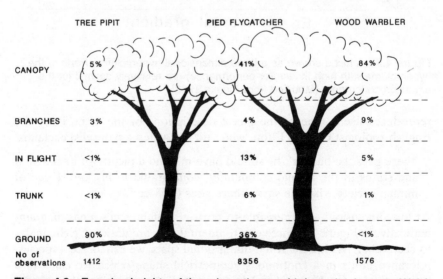

Figure 4.6 Foraging heights of three insectivorous bird species in some Welsh sessile oakwoods. (Modified and redrawn from Stowe 1987)

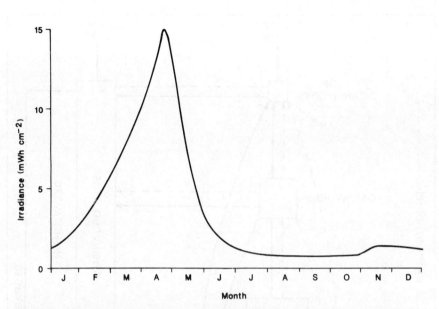

Figure 4.7 Daily total irradiance beneath a deciduous canopy in Madingley Wood, Cambridgeshire. (Redrawn from Peterken 1981)

(*Anemone nemorosa*) and the bluebell (*Endymion non-scriptus*) develop their foliage and flowers early in the spring, well before light levels on the forest floor fall as the trees above develop their canopy (see Figure 4.7).

Ecosystems

An ecosystem is a functionally integrated unit of plants, animals, and their abiotic environment, the integration being provided by flows of energy and the cycling of nutrients (see Figure 4.8). The exchange and release of energy and nutrients comes about when an organism is eaten by another or when it dies. The fundamental process whereby energy enters an ecosystem is through *photosynthesis*, in which green plants absorb radiant energy from the sun in their chloroplast pigments, and use it to fix carbon dioxide into more complex molecules such as sugars, thus storing the trapped energy in chemical bonds. The usual summary equation is:

$$6CO_2 + 12H_2O + 2816 \text{ kJ energy} \rightarrow C_6H_{12}O_6 + 6CO_2 + 6H_2O$$
$$\text{(chlorophyll)}$$

Since only green plants can utilize the sun's energy directly they are said to be within the *autotrophic* subsystem. The term autotrophic literally means self-nourishing. In turn, the energy in green plants is utilized by

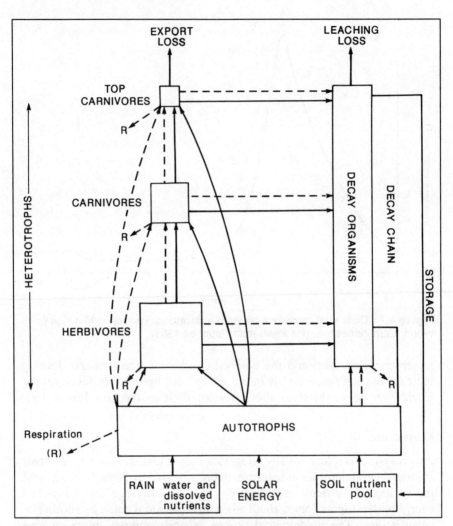

Figure 4.8 Simplified model of the flow energy and nutrients through an ecosystem. Solid lines represent nutrient flows and dashed lines energy flows. (Modified and redrawn from Woodwell 1970)

members of the *herbivore* or *grazing* subsystem. On death, members of both subsystems are the victims of the *decomposer* subsystem, whose activities release elements in a mineral form suitable for re-utilization. A general term for an animal which obtains its energy by eating other organisms is a heterotroph.

The daily amount of solar radiation reaching the British Isles ranges from 10 MJ m^{-2} in southern England to 8 MJ m^{-2} in northern Scotland, corresponding to an annual total of about 3.3 GJ m^{-2}. Of this total only 45 to 50 per cent is photosynthetically active radiation (PAR), defined as the visible part of the energy spectrum between 350–700 nm. In the marine environment, most red wavelengths are rapidly absorbed – after which there is an exponential decrease in PAR with depth. In British coastal waters the PAR at 10 m depth is only 10 per cent of surface levels.

The fate of the PAR once it has been absorbed by a green plant is shown diagrammatically in Figure 4.9. Roughly 25 per cent of the PAR is converted into chemical bond energy, and the rate at which these products accumulate is called *Gross Primary Production (GPP)*. A substantial amount of GPP is used by the plants themselves during respiration so that only about 5–6 per cent of the total radiation energy reaching the plant is net productivity. This accumulation rate is called *Net Primary Productivity (NPP)*. The percentage of GPP lost through respiration varies considerably from ecosystem to ecosystem. Phytoplankton communities respire between 30–40 per cent of their GPP, temperate forests about 50–60 per cent and tropical forests as much as 70–80 per cent.

There is a broad latitudinal trend in NPP ranging from 140 g of dry matter $m^{-2} y^{-1}$ in the tundra to 600–1,200 g $m^{-2} y^{-1}$ in temperate regions,

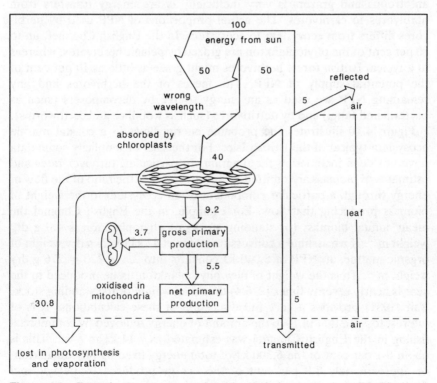

Figure 4.9 Fate of solar energy reaching the leaf of a green plant. (Modified and redrawn from Simpkins and Williams 1981)

such as the British Isles, and 900–2,200 g m^{-2} y^{-1} in the tropics. Inshore waters around the British coast which are enriched by domestic sewage and agricultural runoff attain NPP levels of 200–600 g m^{-2} y^{-1}.

An examination of NPP and climatological data by Leith and Box (1972) has shown that there is a general exponential relationship between NPP in terrestrial systems and evapotranspiration (E) of the form:

$$NPP = 3000 \ (1 - e^{-0.0009695(E-20)})$$

where E is measured in mm, NPP is measured in g m^{-2} y^{-1}, and e is 2.718 (the base of natural logarithms). Annual evapotranspiration in the London region is about 600 mm and when substituted into the above formula results in a predicted NPP of 1,290. In north-east Scotland evapotranspiration is of the order of 300 mm and the predicted NPP is 713.

The NPP determines the potential amount of energy that is available to the consumers in the ecosystem. But consumers actually only consume a proportion of the NPP since they may waste, destroy, or ignore food which might be inaccessible. Of the energy consumed, some is excreted and some lost through respiration. In short, the transfer of energy between the autotrophs and grazers is very inefficient as are energy transfers from herbivores to carnivores. The actual proportions of NPP used by herbivores differs from ecosystem to ecosystem. In the English Channel, up to 80 per cent of the phytoplankton are grazed by pelagic herbivores, whereas in a typical British forest herbivores might graze as little as 10 per cent of the potential supply of NPP. The faeces of the herbivores and any remaining NPP are used as an energy source by decomposers (such as bacteria and fungi) and by detritivores (such as woodlice and earthworms).

Figure 4.10 illustrates the probable energy flows in a coastal marine ecosystem typical of the British Isles. For the sake of simplicity some data have not been included in the diagram. In particular, turnover rates and estimates of biomass are omitted. A turnover rate is the ratio of the flow of energy through a particular component of an ecosystem to the weight of biomass producing that flow. For example, in the English Channel the mean annual biomass (or standing stock) of phytoplankton is 4.0 g dry weight m^{-2}. If we assume a conversion factor of 42 kJ to 2.3 g dry weight of organic matter, an NPP of 5,040 kJ converts into 2.3 × 120 = 276 g dry weight m^{-2}. Thus the weight of new phytoplankton tissue produced in the year is nearly seventy times (276/4) the mean weight of the standing stock. Tait (1981) provides a very full discussion of these calculations. It is of interest to note that in 1976 the amount of energy removed by commercial fishing in the English Channel was estimated as 7.14 kJ m^{-2} y^{-1}. This is about 0.1 per cent of the 6,300 kJ of total energy fixed by phytoplankton, or approximately 8.8 per cent of the energy produced by pelagic and demersal fish (see Figure 4.10).

It is sometimes convenient to summarize the amount of energy, the

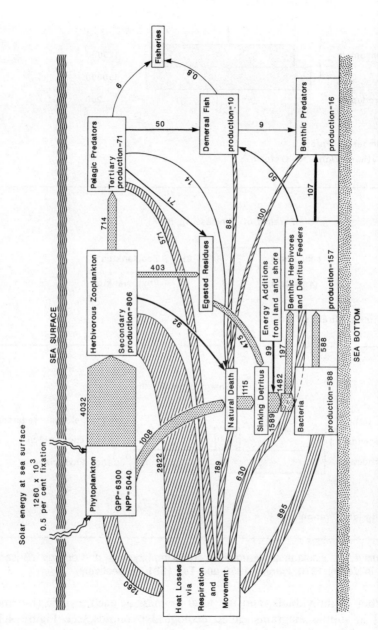

Figure 4.10 Simplified representation of hypothetical energy flows in a marine ecosystem in coastal waters of the British Isles. Units of measurement are KJ m^{-2}yr^{-1}. (Modified and redrawn from Tate 1981)

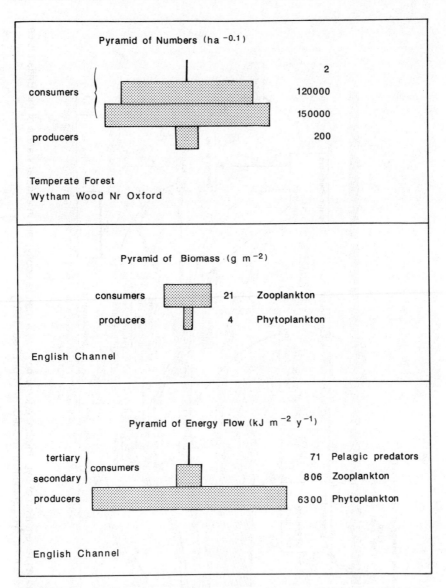

Figure 4.11 Ecological pyramids of numbers, biomass, and energy. (Based on data in Varley 1970; Harvey 1950 and Tait 1981, respectively)

number of individuals, and the total biomass at each trophic (nutrition) level in simple diagrams called ecological pyramids (see Figure 4.11). These types of diagram presuppose that animals can be satisfactorily classified as either carnivores or herbivores. Many animals are opportunistic feeders and may feed on both animal and vegetable matter; such animals are called omnivores. Some animals, such as frogs are herbivorous during their immature stages and carnivorous as adults. Thus the distinction between trophic levels in the real world is often blurred.

A complete food network reveals a complex web-like structure. Simple linear paths through such a *food web* are called *food chains*. Figure 4.12 illustrates a simplified food web for Wytham Woods near Oxford. Small mammals and birds, such as the blue tit (*Parus caeruleus*) have been omitted from the diagram as they occupy several trophic levels. The blue tit, for example, feeds on beech mast (autotrophic matter), females of the winter moth (*Operophthera brumata* – a herbivore) and on spiders and beetles (carnivores).

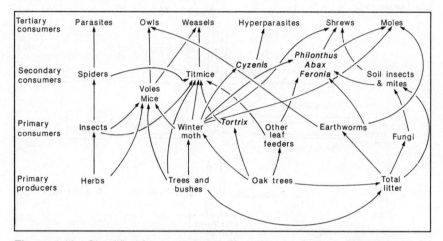

Figure 4.12 Simplified food web for Wytham Woods, Oxford. (Modified and redrawn from Varley 1970)

Unlike energy in an ecosystem, which enters as light and leaves as metabolic heat in a one-way flow, mineral nutrients such as calcium, potassium, and nitrogen can be recycled within the system. Such cycles are sometimes called *biogeochemical cycles*. Pools of nutrients are stored in components of the ecosystem which may be either biotic or abiotic. Over time, the total flow of any one nutrient does not vary much because of self-regulating negative feedback loops. These either release or store elements, depending on whether or not there is a deficiency or excess in the system. Within a typical woodland ecosystem the major nutrient cycle occurs between the soil and green plants (see Figure 4.13). Minerals are absorbed by plant roots, and are incorporated into wood and seasonal tissue such as leaves. When this tissue dies, minerals are released back into the soil by leaching and the activities of decomposers. In general there is only a very slow loss of mineral nutrients from an undisturbed ecosystem. However, if it is cropped in any way by the removal of plants or animals,

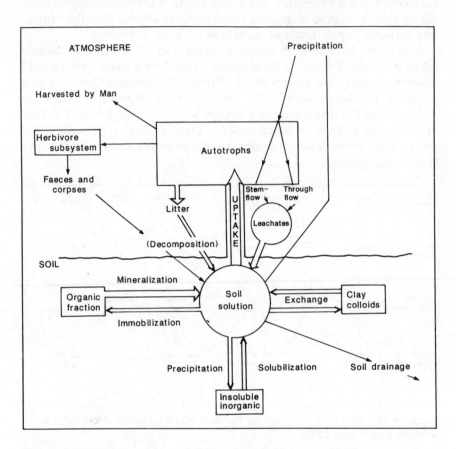

Figure 4.13 Major pathways of nutrient transfer in a woodland ecosystem. (Modified and redrawn from Packham and Harding 1982)

as in commercial farming systems, losses may be high and lead to soil infertility.

At the International Biological Program site at Meathop Wood, a mixed sessile oak, ash, and birch woodland in south Cumbria, nutrient contents and flows have been measured in great detail. Major nutrients are translocated from the soil into woody tissues and especially to growing leaves and shoots which commonly contain more than ten times as much nutrient as the wood in the same species. Levels of leaf nutrients vary through the growing season and some, such as nitrogen and phosphorus, may be drawn back down into the woody tissue before leaf senescence. Nutrients in the ecosystem are also cycled locally in the ground flora, and because both trees and ground flora are subject to leaching by rain, some chemicals may leak out of the system. But at Meathop, nitrogen and

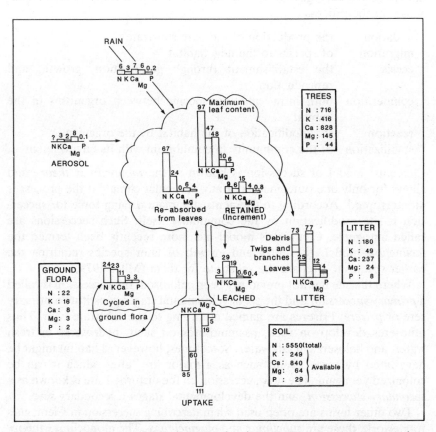

Figure 4.14 Mean nutrient flows (kg ha^{-1} yr^{-1}) and nutrient contents (kg ha^{-1}; in boxes) in Meathop Wood, south Cumbria, a mixed oak, ash, and birch woodland developed on brown earths. (Redrawn and modified from Packham and Harding 1982)

phosphorus are apparently tightly cycled (see Figure 4.14), with very low losses from the canopy and the soil (Brown 1974).

Succession

Areas of bare substrate, both on land and in the sea, are rapidly colonized by plants and animals. Over time, the habitat becomes modified by the invading species (through such processes as organic enrichment), and more demanding, competitive species enter. The process of changes in species composition over time is called *succession*. The last stage in a succession is called the *climax* and is often idealized as a community having constant species composition. The classical concept of succession was formulated by the American ecologist, F. E. Clements in 1916 who recognized six main phases in the process:

nudation   the production of the bare substrate
migration   of species to the new habitat
ecesis   the establishment through germination, growth, and reproduction
competition   both intra- and interspecific between organisms in the habitat
reaction   the modification of the habitat by the organism
stabilization   of the community in equilibrium with its environment

Clements' model of succession is known as the *monoclimax theory* and allows for only one outcome under any particular climate if the process is uninterrupted. According to Clements, the main driving force for succession is site modification by the community itself. Such successions are called autogenic. Clements's model has more recently been termed the *facilitation* model, the entry and growth of later species requiring the earlier species to prepare the substrate for them (Miles 1979).

Where the site has not previously been colonized, the succession is called a *primary succession*,and the six developmental stages constitute a primary sere or *prisere*. Priseres are named according to the substrate type. Thus lithoseres develop on rock, psammoseres on sand, hydroseres in fresh water, and haloseres in salt water. Sometimes, however, a habitat might be devastated by some event, such as a major fire, after which it can be colonized yet again. The new succession on the disturbed site is known as a *secondary succession*, and the developmental stages a secondary sere.

Two other terms are often used when describing succession in Clements's framework, these are *subclimax* and *plagioclimax*. The monoclimax theory argues that a particular succession of species occurs in each climatic zone, and in addition, that this succession may be stopped, either temporarily or permanently, by so-called *arresting factors*, which might be geomorphic, edaphic, or perhaps biotic in character. A succession which is arrested in its development is said to have reached a subclimax stage. In the British Isles, alder (*Alnus glutinosa*) woodlands, also known as carr woodlands, are frequently the subclimax vegetation on gley soils subjected to repeated flooding. This species will remain dominant, and the vegetation will remain stable, until such time as the arresting factor is removed. If, for example, a farmer drains adjacent fields, or perhaps a river changes its course, the water table will be lowered and the woodland soils will dry out. On these dryer soils, a subseral succession will then lead to the climatic climax vegetation, which in lowland Britain is a mixed oak forest. When man or his activities are the arresting factors, the terms plagioclimax and plagiosere are used (Greek – *plagios* meaning oblique or deflected). Heather moorlands are common in the uplands of the British Isles and occupy sites which were formerly wooded. Today, regular burning of heather moor-lands of northern England and Scotland keeps the community in the plagioclimax state. Without management, the heather community would grow tall and degenerate in 20 years or so. Eventually it would be replaced by birch scrub.

Some ecologists, such as Tansley, have argued that climate is not all dominant and instead the vegetation will stabilize into edaphic, physio-graphic, biotic, or anthropogenic climaxes. This theory is known as the *polyclimax theory* of succession. It is, perhaps, important to emphasize that both models of succession are theories, and they have not been tested very thoroughly. One remarkable study of 159 post-glacial (Holocene) hydro-seres in Britain by Walker (1970) used the stratigraphic evidence from topogenous mires to show that far from woodland being the natural culmination of the succession, as predicted by Tansley, the likely outcome is *Sphagnum* bog (see Figure 4.15).

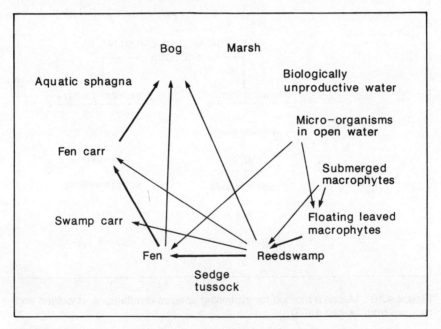

Figure 4.15 Common hydrosere transitions deduced redrawn from pollen analytical data (Modified and redrawn from Walker 1970)

The process of succession occurs more rapidly in marine benthic communities than in terrestrial systems because the life-span of the dominant species is shorter. One or two seaweed species, such as the knotted wrack (*Ascophyllum nodosum*), have a life-span of up to 25 years or so but most are much shorter lived.

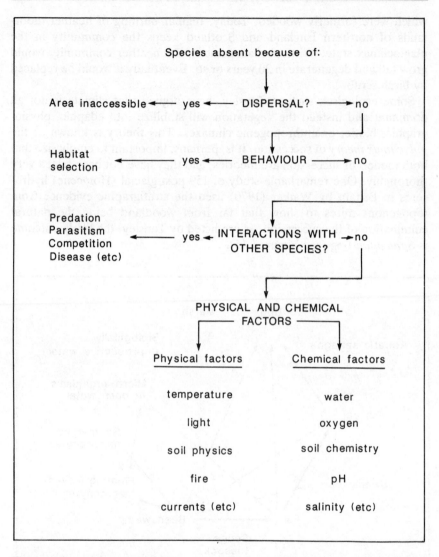

Figure 4.16 Macan's method for explaining species distributions. (Modified and redrawn from Macan 1963)

Explaining distributions

Since the study of species distributions is central to biogeography it is useful to have a model in which explanation proceeds in an orderly fashion. One such procedure, suggested by Macan (1963), is illustrated in Figure 4.16. To explain the absence of a species from an area, we enter the top of the diagram and proceed down the chain. It may seem rather strange to examine absence rather than presence since this seems to run counter to the scientific method. None the less, Macan's suggestion is a useful heuristic device. Incidentally, the possible effects of man on distributions is easily accommodated, since man may be regarded as an 'other species'. In the next two chapters we shall examine various parts of Macan's model in detail.

Chapter five

Gaining a foothold

The distribution of an organism does not totally depend on its ability to tolerate extrinsic environmental factors such as temperature, humidity, or salinity. The absence of an organism from an area might be due to competition, poor dispersal capability, or perhaps because of some behavioural mechanism such as habitat selection. In this chapter, we shall examine the biogeographical implications of these so-called intrinsic limiting factors (Udvardy 1969).

Dispersal and migration

It is not at all easy to find an agreed definition of the terms dispersal and migration. Plant ecologists, entomologists, ornithologists, and biogeographers frequently use the terms in different contexts and sometimes as if they meant one and the same thing. It is probably impossible to satisfy all biologists and biogeographers, but the definitions given by Taylor (1986) are very interesting. *Dispersal* simply means separation and becoming further apart and, by implication, an individual cannot disperse by itself but only from other members of the population. It also follows that dispersal of an individual has no reference to place and is only part of some collective behaviour. In contrast, *migration* relates to geographical co-ordinates, not to individuals, although, as Taylor notes, it may be done in company or alone. The initial co-ordinates of the migration are the individual's birthplace. The terminal co-ordinates for a completed migration are the birthplace of the next generation.

This description of migration leads to three types. *Random dispersive migration* is a one-way involuntary emigration in which the organisms have no control over the immigration end-point. The outcome is a random distribution which may or may not lead to dispersion. A second type is known as *dynamic migration* which is a one-way actively initiated migration which is under control during emigration, the passage to the new location, and during immigration. Taylor's third type is called *homeostatic* migration, and is an actively controlled two-way migration in which navigation to the location is essential.

Since there is no behavioural mechanism at work in the dissemination of plant spores and seeds, the process is properly one of dispersion. For the most part, the disseminules (spores, seeds, and fruits) disperse in air and water currents, although sometimes animals (and even man) may be involved. Many flowering plants have seeds and fruits which are winged or plumed to reduce their terminal velocity, and, in general, dispersal is enhanced by increasing the release height. A few examples of structures modified for wind transport are shown in Figure 5.1. Because dispersal by wind is relatively inefficient, in the sense that many seeds will fall on unsuitable substrates, wind-dispersed seeds tend to be produced in great abundance and hence are often small. The wide dispersal of very small seeds means that they are often the first to arrive at newly exposed sites. During the days of the London Blitz, from September 1940 to May 1941, approximately one-third of the City of London was destroyed and a great many bombed sites were colonized by wind-dispersed weedy species. A survey by Salisbury in 1942 revealed that some 27 plant species had dispersed into these sites and by the end of the war the figure had risen to 126.

R. L. Praeger investigated the buoyancy of British plant seeds and fruits, and found that a great many float, at least for a short while, and sometimes longer when aided by adaptations – such as those on the fruits of the curled dock (*Rumex crispus*) and the arrow-head (*Sagittaria sagittifolia*) which have corky wings to aid buoyancy. On the west coast of Ireland, tropical fruits and seeds regularly get washed ashore having drifted in the Gulf Stream perhaps for as long as a year (the record crossing for a drift bottle from Hispaniola in the Caribbean to south-west Ireland is 337 days – a journey of about 6,600 km). Some disseminules reaching Irish beaches are those of so-called *refuse* species, which have been thrown overboard in the garbage from passing ships. Others, such as the large, dark maroon-brown, heart-shaped seeds of the woody vine *Entada gigas*, are true peregrine species probably having drifted in ocean currents all the way from the West Indies (Nelson 1978). Several species are still capable of germination in spite of their ocean journey, though they cannot survive outdoors in Ireland.

One amazing dispersal that led to successful germination was the discovery, in 1971, of *Juncus planifolius*, on the south side of Lough Truscan in remote west Galway. The nearest known stations for this rush species are in the Southern Hemisphere some 12,000 km away! Although there remains the possibility that the seed was introduced by man, this seems very unlikely and it is more probable that it was carried by migrating birds. Some mosses produce incredible numbers of spores, of the order of 10–25 µm in diameter, which readily become airborne in a light wind. That

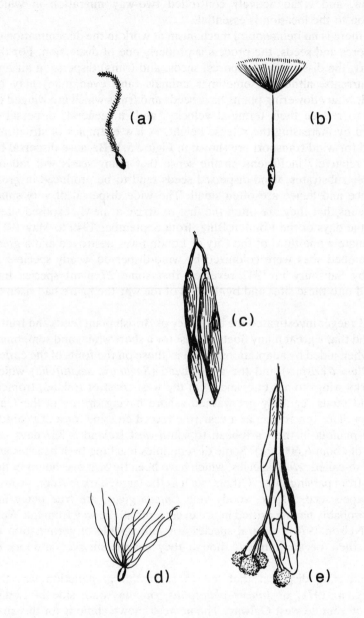

Figure 5.1 Examples of wind-dispersed seeds and fruits: (a) old man's beard (*Clematis vitalba*) – a fruit with a feathery style; (b) common dandelion (*Taraxacum officinale*) – a hairy, plumed fruit; (c) ash (*Fraxinus excelsior*) – two winged fruits; (d) downy, plumed seed of rose-bay willow-herb (*Chamaenerion angustifolium*); (e) Lime (*Tilia* spp.) – fruits suspended from a strap-like bract.

moss spores can be transported at least 1,000 km is attested by the discovery of *Tortula* spp. and *Funaria* spp. in a newly formed volcanic crater on Deception Island, Antarctica within 9 months of the eruption. Long-distance dispersal across large barriers, such as oceans and mountain ranges, is sometimes called *waif dispersal*.

To what extent the effectiveness of dispersal determines geographical distribution is an important question which has been little studied by biogeographers. In his classic monograph on the East Anglian flora, Salisbury (1932) discussed the spread of recent immigrants, and directly addressed the point as to whether or not the rate of spread is related to the type of dispersal mechanism. The Canadian fleabane (*Erigeron canadensis*) has seeds covered in down which are easily dispersed by wind. However, its spread from London, where it was first recorded in 1690, has been relatively slow (see Figure 5.2). In contrast, the monkey-flower (*Mimulus guttatus*) which has no particular dispersal mechanism has spread over most of the British Isles since it was first recorded in Wiltshire in 1832 (see Figure 3.4). Salisbury came to the conclusion that the diverse mechanisms of dispersal are probably less important that other environmental factors in explaining plant distributions in the British Isles. In a later study, he refined this argument by comparing the dispersal efficiency of four species of poppy (*Papaver* spp.). By so doing, environmental factors which might influence distributions could be kept constant since he presumed they had similar ranges of tolerance. He found that widespread species, such as the field poppy (*Papaver rhoeas*) produced many more seeds than more local species, such as the round prickly-headed poppy (*Papaver hybridum*). He also noted that widespread species carried their seed capsule nearly twice as high above the ground, compared with more local species. In poppies, at least, dispersal efficiency does seem to be an important factor in determining their range.

According to Taylor, migration in animals is, in general, a process of optimization in the spatial environment by selective movement, not by reproduction: improved reproduction is the result of good site selection not the cause of it. Taylor's first class of migration, random dispersive, does not quite fit in with his notions of selective movement, but this merely illustrates the problems of classifying migrations. Small insects are often caught up in strong weather systems and blown hundreds of miles from their birthplace.

A typical example of random dispersive behaviour was the large immigration of the grey and diamond-back smudge moth (*Plutella maculipennis*) along the coasts of eastern and north-eastern Britain in June 1958. This moth is indigenous, and is a pest of cabbage and lettuce crops, but large increases in its population are nearly always associated with the addition of migrant populations. An inspection of the weather charts for this date shows that the migrants were probably blown in from the shores

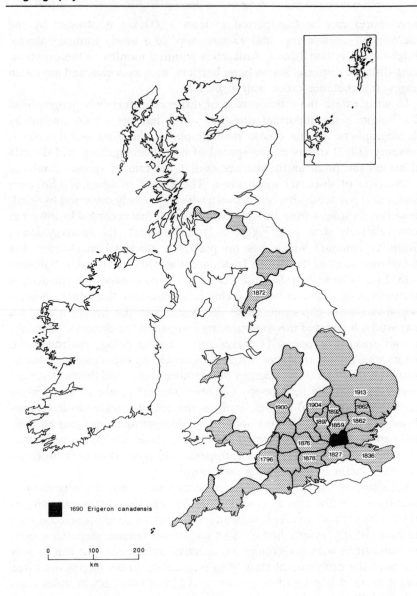

Figure 5.2 The dispersal of the Canadian fleabane (*Erigeron canadensis*). (Modified and redrawn from Salisbury 1932)

of the Baltic on a strong easterly air-stream. This air-stream also blew part of the migrating population way out into the Atlantic Ocean, where it was observed by a weather ship south of Iceland.

In some insect orders such as the Lepidoptera (butterflies and moths), dwarf races have evolved as an adaptation to avoid being blown away by the wind, especially in coastal habitats. On the exposed limestone pastures of Great Orme's Head, in North Wales, the small brown butterfly known as the grayling (*Hipparchia semele*) is represented by the dwarf race *thyone*.

Random dispersive migration is one mode of spread of pathogenic viruses. Foot-and-mouth viruses are now suspected of being spread by wind which picks them up as diseased animals exhale. Once suspended in the wind there is some evidence of successful transmission over distances up to 150 km. The March 1981 outbreak of foot-and-mouth on the Isle of Wight was attributed to wind-borne migration of viruses from the Continent.

In marine environments planktonic larvae drift in water currents and can migrate successfully if the characteristics of the water body or substrate is suitable. The barnacle *Elminius modestus* is a native of southern and eastern Australia and New Zealand, and arrived in the English Channel sometime during the Second World War, having been carried from its native haunts on the hull of a ship. It was first seen in Chichester harbour in 1944, and from measurements made over the following ten years, *Elminius* was shown to have migrated at a rate of about 30 km y»fb[1] (Crisp 1958). The rapidity of the migration of *Elminius* was undoubtledly helped by long-shore currents wafting larvae on to suitable rocky habitats. The dispersive migration of *Elminius* away from its first Irish Sea location on the coast of Morecambe Bay is shown in Figure 5.3.

One-way dynamic migration is typical of many invertebrates, such as aphids and moths. In such species, migrations are often compulsive and pre-determined. The individual has no choice, being inhibited from reproduction without first undergoing flight. Much knowledge about insect migration has been learnt from the detailed mapping conducted as part of the Rothamsted Insect Survey and collaborative European schemes such as EURAPHID (Taylor 1986).

In Taylor's third class of migration – the homeostatic type – the return to the breeding site and its location through navigation are essential features. The seasonal return flight of migrant birds is an obvious example. Homeostatic migration is also common in fish such as the mackerel (*Scomber scombrus*) and herring (*Clupea harengus*). The circuit of migration runs from spawning ground to nursery ground, from nursery ground to feeding ground, and from feeding ground back to spawning ground. Spawning grounds tend to be fixed both in time and space, and migratory fish – such as the herring – return year after year and may be thought of as navigating to the site. From the spawning ground, fish larvae drift away on regular currents towards the nursery grounds, which are often inshore, away from deep-water predators. As the little fish grow, they move into deeper water feeding grounds and on reaching sexual maturity, return to join the adult spawning shoals.

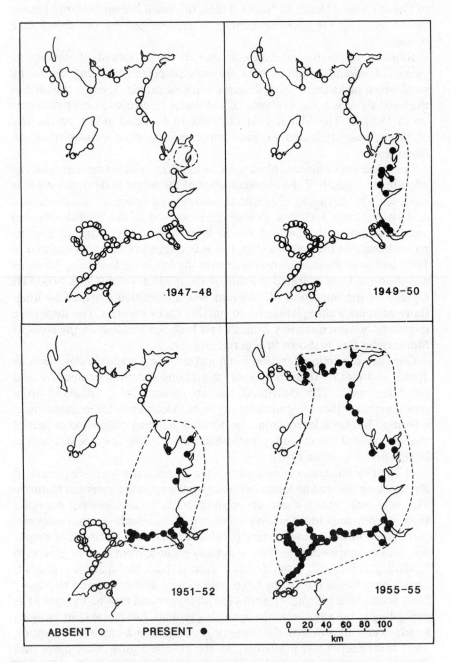

Figure 5.3 The random dispersive migration of the barnacle (*Elminius modestus*) in the Irish Sea. (Modified and redrawn from Crisp 1958)

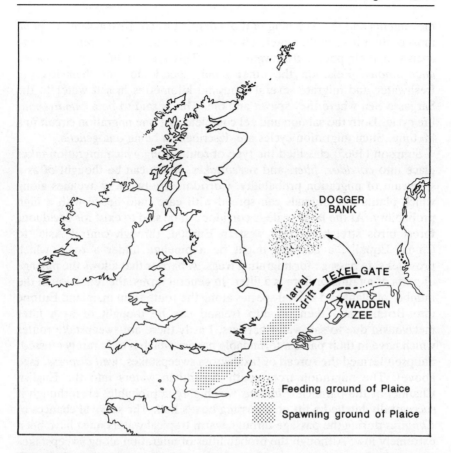

Figure 5.4 Migration circuit of the southern North Sea plaice stock (*Pleuronectes platessa*).

Plaice (*Pleuronectes platessa*) belonging to the southern North Sea stock complete a migration circuit of about 300 km. Spawning takes place in the Southern Bight between the Thames and the Rhine and the larvae drift down current in the most saline and clear water toward the so-called Texel Gate off northern Holland (Figure 5.4). The Texel Gate is a hydrographic gate at the junction of two water bodies, one being homothermal English Channel water, the other, stratified Dutch coastal water. Tides running parallel to this boundary produce bottom water which moves inshore and carries the little plaice into the Dutch Wadden Zee. As they mature they move out into the feeding grounds of the Dogger Bank from where they make annual migrations back to the spawning area.

Apparently, some fish, such as the European eel (*Anguilla anguilla*) and the salmon (*Salmo salar*) are forced to migrate long distances, as their breeding and feeding grounds have been separated by plate tectonic movements and the widening of the Atlantic Ocean. British salmon spend most of their lives in the waters off southern Greenland, and return once to spawn in their parent freshwater river. This type of life cycle is called *anadromous*. Eels, on the other hand, spend most of their lives in freshwater and migrate several thousand kilometres in salt water to the Sargasso Sea where they spawn and die. This is said to be a *catadromous* life cycle. Both the salmon and eel complete just one migration circuit in a lifetime. Such migration cycles are described as being *ontogenetic*.

Simpson (1965) classified the type of route along which migration takes place into *corridor, filter,* and *sweepstakes*, which can be thought of as a spectrum of migration probability. Corridors are two-way avenues along which plants and animals can spread with ease, and hence with a high probability. At the large scale, a corridor can be said to exist for deciduous forest birds stretching from western Europe, through central Asia, to China. Equally, a corridor might be a pipeline under a road, which provides safe passage for migrating frogs. A barrier that allows the passage of parts of a fauna or flora is a filter. In general terms, the reduction in the number of plant and animal species along the route from mainland Europe into Britain, and then on into Ireland can be thought of as a filter mechanism due to salt-water barriers. Lastly there are sweepstake routes which have in their paths a formidable barrier which is only rarely crossed. Simpson termed the spread of taxa across sweepstakes, *waif dispersal* (see above). The migration from south Australian waters into the English Channel of the barnacle *Elminius* was highly improbable, even though it successfully hitched a lift on a passing boat's hull. The survival chances of *Elminius* during the passage through warm tropical waters must have been extremely low. Although the probabilities of migration along sweepstakes are small, Simpson thought that migration would take place if enough attempts were made. Clearly, there is a very large population of barnacles in Australian waters, and given the passage of time, even this rare migration event was successful. It should be noted that the same route might be a filter for one taxon and a sweepstake for another because of differential capacities for migration.

Behaviour

A species hardly ever occupies its potential range as defined by its tolerances to environmental factors and may show an active preference for particular habitat types. The behavioural mechanisms for habitat selection are complex and not well understood. This is partly because we must try to understand habitat choice through the 'eyes' of the organism and not through our own perceptions of the habitat.

Habitat selection is well known in a number of bird species such as the nightingale (*Luscinia megarhynchos*). This bird is a summer migrant and is

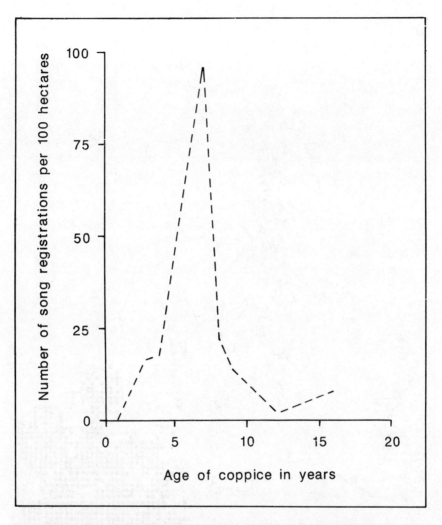

Figure 5.5 The relationship between the number of nightingale song registrations and age of coppice woodland. (Modified and redrawn from Stuttard and Williamson 1971)

usually found in woods, overgrown hedgerows, and thickets, where it skulks and forages amongst the undergrowth. Nightingales show a particular preference for a woodland structure produced as a result of coppicing. In Britain, the coppice management of woodlands has a history stretching back at least 1,000 years, and much remaining woodland has been coppiced at one time or other; the practice seems not to have occurred in Ireland (Rackam 1976). Coppicing involves cutting all the woody growth and allowing the new coppice shoots to spring up from the

(a)

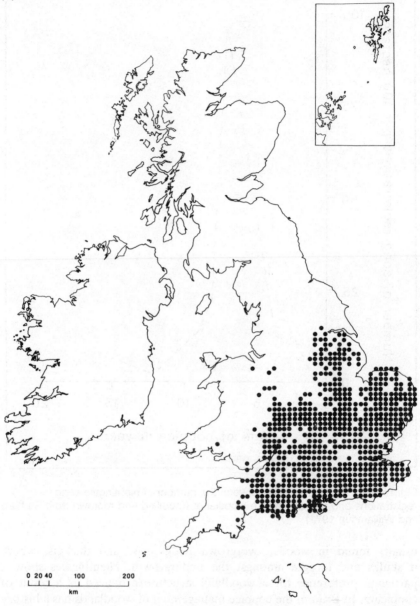

Figure 5.6 The distribution of (a) nightingales; (b) recent coppicing. Both maps based on 10 km² grid sampling. (Redrawn and modified from Sharrock 1976 and Peterken 1981)

(b)

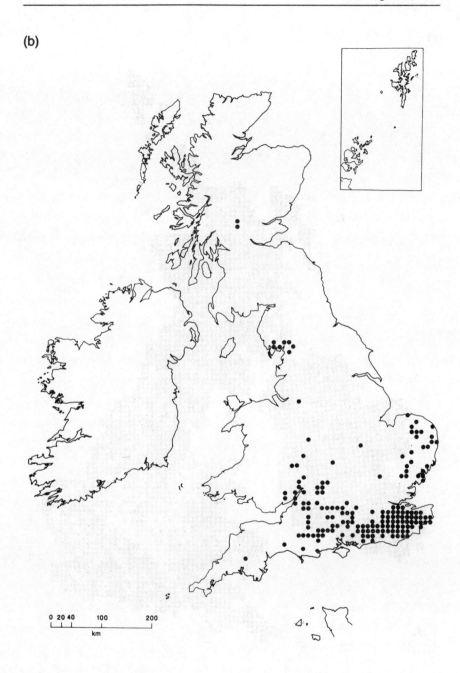

Figure 5.6 continued

(a)

Figure 5.7 The distribution of (a) meadow pipit; (b) tree pipit. (Reproduced from Sharrock 1976)

(b)

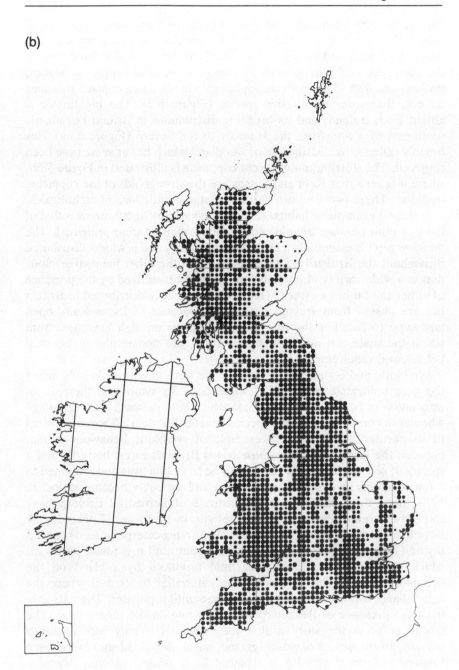

Figure 5.7 continued

stumps to form small, straight poles which are harvested every 5 to 30 years. The resulting woodland structure provides secretive nesting sites near the ground, and the light and warmth reaching the woodland floor in the early years of pole growth encourage a plentiful supply of beetles, spiders, and flies – on which the nightingale feeds. Long coppice rotations do not, therefore, favour this species (Figure 5.5). The nightingale is absent from Ireland, and its breeding distribution in Britain lies mainly south-east of a line from the Humber to the Severn (Figure 5.6a) This broadly reflects the distribution of woodland which has at some time been coppiced. The distribution of recent coppicing is illustrated in Figure 5.6b, where it is seen that Kent and Sussex are the strongholds of the coppicing tradition. These two counties also have large populations of nightingales.

A second example of habitat selection influencing distributions is that of the tree pipit (*Anthus trivialis*) and meadow pipit (*Anthus pratensis*). The meadow pipit is essentially a bird of open country and is widely distributed throughout the British Isles (Figure 5.7a) On the other hand, tree pipits nest in a wide variety of habitats but all are characterized by the presence of either tall bushes or trees. Tree pipits are widely distributed in Britain but are absent from Ireland presumably because of its basically open landscape (5.7b). For the tree pipit, song-posts on high branches, from which the male can advertise his territory, are apparently an essential behavioural requirement.

Symbiotic and commensal behaviour are both important in influencing the geographical distribution of some taxa. In symbiosis, there is an association of two different species in which both partners gain advantage whereas in commensalism one species benefits but does no apparent harm to its partner. A rather nice example of symbiotic behaviour is that between the large blue (*Maculinea arion*) Britain's rarest butterfly and a species of red ant (*Myrmica sabuleti*). The large blue was once confined to a few localities in Cornwall and Devon and probably became extinct in 1979. Since then, scientists at the Institute of Terrestrial Ecology have successfully reintroduced it at a secret site in south-west England. The large blue is totally dependent on ants. The young caterpillar feeds on wild thyme (*Thymus drucei*) but later crawls about until it is found by an ant which is attracted by the sugary fluid produced by a gland on the caterpillar's back. The ant carries the caterpillar to its nest where the caterpillar grows fat on a diet of ant larvae until it pupates. The ants gain from the presence of the caterpillar by feeding on the sugary fluid. The crucial point in the story is that the symbiosis is only with *Myrmica sabuleti*, an ant species of closely grazed, warm, downland sites in southern England. *Myrmica sabuleti* is replaced by a related species, *Myrmica scabrinodis* if the grassland sward is allowed to grow tall. Alas, *Myrmica scabrinodis* destroys the grubs of the large blue (Nature Conservancy Council 1986)

Habitat selection by some bird species varies according to their overall population levels. For example, populations of wrens (*Troglodytes troglodytes*) fell during the severe winter of 1962/3 and numbers rose sequentially in different habitats. At its lowest ebb, the wren populations were mainly concentrated in woodland and stream-side habitats. As numbers increased, these habitats became saturated, more gardens were occupied and, finally, wrens began to move into farmland hedgerows. This shift in habitat distribution suggests that gardens and field hedgerows are sub-optimal habitats for this species.

Competition and predation

The establishment of a geographical range is not only contingent on successful dispersal or migration into suitable habitats but also on success in any competition and the avoidance of predators. As was noted in Chapter four, competition between species is likely to be most severe where there is a high degree of niche overlap and in such a situation competitive exclusion may well take place.

Competition, and also the lack of it, is well illustrated in the biogeography of the two species of rat, the black rat (*Rattus rattus*) and the brown rat (*Rattus norvegicus*). The black rat, or ship rat as it is sometimes known, was formerly distributed in south-east Asia and the Middle East. It was accidentally introduced into the British Isles in the late twelfth century, in, as tradition would have it, the baggage of returning Crusaders (Lever 1979). Its presence in Ireland was mentioned by Giraldus Cambrensis in 1185. The black rat also brought with it the flea *Xenopsylla cheopis*, the vector of the bubonic plague. In the next couple of centuries the black rat became widespread in the British Isles easily competing with indigenous rodents, such as voles and mice. The brown rat arrived in England in 1728 or 1729 and was reported in the Dublin district about 1722. It is both larger and more pugnacious than the black rat, and its ferocity led to numerous horror stories in the press of the day (Fairley 1984). A statement in *Walsh's Impartial Newlsetter* dated 1729 reads:

> People killed several as big as Katts and Rabbits. This part of the country is infested with them. Likewise we hear from Rathfarnham that the like vermin destroyed a little girl in the Field; they are to be seen like Rabbits, and are so impudent that they suck cows.

The black rat is clearly no match for the brown rat which, at about 0.5 kg, is about twice the weight of the former which has, with some notable exceptions, driven the smaller species out of most habitats. The success of the brown rat is due to a number of factors such as its ferocity, its swimming skills, high fecundity (up to seven litters a year), omnivorous habits, and its use of subterranean habitats, such as sewers and drains

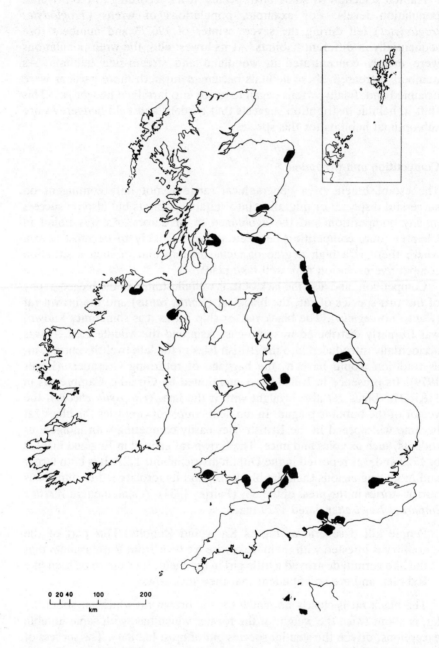

Figure 5.8 The distribution of the black rat in the British Isles. (Modified and redrawn from Lever 1979)

where it can escape persecution from man. In Ireland the brown rat had become the most common species within about 20 years of its entry into the country, and by the middle of the nineteenth century was also widespread in Britain. In the face of such competition, the range of the black rat has collapsed into isolated pockets around major ports and dockyards (Figure 5.8), where it seems to have established permanent colonies.

A similar picture of competitive exclusion is sometimes described for the interactions between the indigenous red squirrel (*Sciurus vulgaris*) and the introduced grey squirrel (*Sciurus carolinensis*). Indeed, conservationists often cite the possibility of competition with our native fauna and flora as one possible danger in the careless introduction of foreign species to the British Isles. Of course, not all competition between species takes place on such a large scale as that of the brown and black rat. In plants, competitive reactions tend to be more localized. In the nearshore zone, for example, the regular zonation of fucoid seaweeds is, in part, due to the intense competition for light, since nutrients are generally well dispersed in water and not an important factor.

If competition is thought of as interaction within a trophic level, then predation represents the interaction between trophic levels. Predation and competition interactions can also be related, as in the case of two competing species that are reduced in numbers by their predators to such an extent that competition is prevented. Grazing animals are predators on plants, and the effects of increases and decreases of grazing pressure has a considerable influence on species distributions. For example, on the chalk grasslands of southern England tall grasses such as the oat-grass (*Arrhenatherum elatius*) only invade habitats after sheep or rabbit grazing has ceased.

In their studies of the common mussel (*Mytilus edulis*) on the coast of southern Ireland, Kitching and Ebling (1967) found that one of the main factors influencing its location was the distribution of predatory crabs. On exposed parts of the coast, where wave energy and currents are strong, mussels are plentiful, due to the absence of crabs in these environments. In more sheltered inlets, crabs are more common and only large mussels attached to vertical surfaces are free from predation.

Migrations, extinctions, and equilibria

A glance at a species list for a given taxon generally reveals a relationship between species numbers and island size. Great Britain, for example, has forty-four indigenous terrestrial mammal species, while Ireland has only twenty-two. It might be thought that this is just the effect of a salt-water barrier, but the reduction in species numbers also affects bats – which can easily fly the required distance (Gorman 1979). An obvious reason which accounts for fewer species on smaller islands is island size but, as ecologists discovered many years ago, there was a consistent type of relationship between size and number. In many parts of the world it has been noticed that a tenfold increase in island area is associated with a doubling of the number of species present for any given taxon. Data such as these can be plotted on a graph and the relationship between island size and species number represented by the power function:

$$S = CA^z \qquad [1]$$

where S is the number of species, C is the intercept, A the area of the island and z is the slope of the curve. In practice, data are collected from field surveys and published material, plotted on a graph and Equation [1] fitted as a linear regression model. Technically speaking Equation [1] is linear in the parameters, and for estimation purposes can be transformed to the logarithmic form (it does not really matter whether we use natural or common logarithms for the transformation):

$$\ln S = \ln C + z \ln A$$

Islands can be either true islands surrounded by water or surrogate islands such as woodland patches in a 'sea' of grassland. Roden (1979) studied the vascular flora of 16 islands in Loch Corrib, near Galway. The approximate area of each island was calculated from a 1 : 10,000 map and the total number of different plant species obtained by field survey. The data are shown in Table 5.1.

Table 5.1 Species-area data for sixteen islands in Loch Corrib, Galway

Number	Name	Area (m²)	No. of species
1	Carrickaslin	4,850	48
2	Coad Island	16,800	76
3	Bilberry Island	14,600	74
4	Bronteen Island	11,500	70
5	Flag Island	600	41
6	Mark's Island	2,530	49
7	Nameless	600	32
8	Clydagh Island	500	33
9	Inish Cairbre	11,860	115
10	Long Island South	2,935	60
11	Nameless	500	46
12	Illaunacreeva	600	47
13	Lackadunna Rock	3,240	64
14	Goat Island	2,800	55
15	Illaunnaneel West	32,400	98
16	Illaunmahon	17,300	72

Source: Roden (1979)

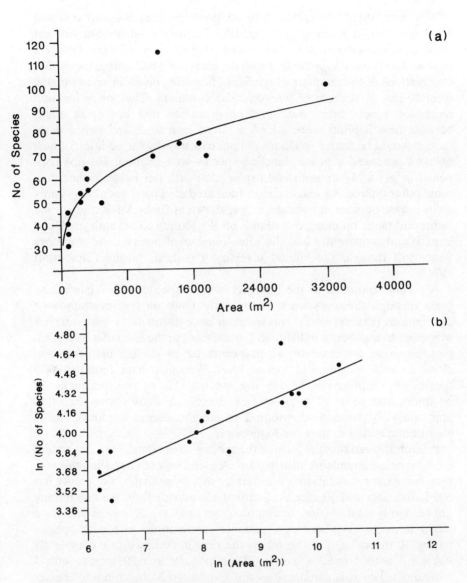

Figure 5.9 Species-area plots for vascular plants on islands in Loch Corrib, Galway (a) Arithmetic scales; (b) logarithmic scales. (Data from Roden 1979)

Figure 5.9a shows the untransformed data plotted on arithmetic scales. In Figure 5.9b the logarithmically transformed data are plotted, together with the estimated species–area regression line:

$$\ln (\text{no. of species}) = 2.318 + 0.2143 \ln (\text{area})$$
$$(t = 7.012)$$

121

Why there should be such a clear relationship between island area and floral and faunal richness is a question frequently asked, but not yet satisfactorily answered. One explanation suggested by Williams (1943) is the *habitat diversity hypothesis*. From the literature, Williams gathered 227 observations of the numbers of species of flowering plants in areas ranging from 10 cm^2 to the size of the American continent. The power function regression model fitted well, and he presumed that this came about because new habitats were added as areas increased, and hence more species would be found. Williams did not explain why the addition of new habitats produced a power function species–area relation. He just happened to get a higher statistical explanation with this model rather than some other type of regression curve. Increased area need not always imply an increased number of habitats, as was shown in Simberloff's study of the arthropod fauna on mangrove islands off the Florida coast; both small and large islands apparently had the same number of habitats and therefore Simberloff thought the habitat diversity hypothesis invalid (Simberloff 1976).

A second approach to the interpretation of species–area relationships came through Preston's work in the early 1960s on species–abundance distributions (Preston 1962). Species–abundance distributions relate number of species to number of individuals (abundance) in the form of a histogram plot. Suppose, for example, we had captured insects in a light trap and classified each specimen at species level. We might have found that 35 species were represented by only one specimen, 28 by two specimens, 18 by three, and so on. Typical species–abundance distributions are often approximately normally-distributed when abundance is plotted on a logarithmic scale i.e. they are lognormal.

Preston showed that the form of the species–area relation was dependent on the species–abundance distribution. A useful way of visualizing this is to take two extreme situations. We start by supposing that a community has 300 individuals and 30 species, and that all individuals in the community are randomly located (the random placement model). At one extreme, we might have an even species–abundance distribution, with each species having 10 individuals; at the other, the distribution might be completely uneven – with 1 species having 271 individuals and 29 species with 1 individual. The two communities are now sampled by counting species numbers in progressively larger quadrats (square sampling frames). For the *even* distribution we expect a species–area plot to increase rapidly and then plateau, because there are no new species to be found as a result of increasing the size of quadrat. For the *uneven* model, the species–area curve is linear because there are so many individuals of the same species, which obviously slows down the rate of capture of new species as the sampling area increases. In Figure 5.10, the species–area curves for these two extreme situations are shown together with a power function which

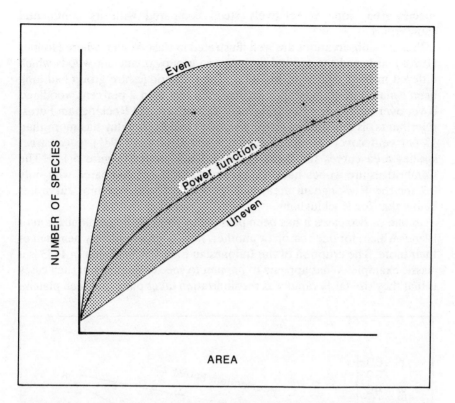

Figure 5.10 Graph showing possible species-area curves under the hypothesis of random placement for communities with 30 species and 300 individuals. An *even* community has ten individuals per species; and an *uneven* community has one species with 271 individuals and 20 species with one individual. The power function curve is shown for comparison. (Modified and redrawn from McGuinness 1984)

Preston showed would be derived if the species–abundance distribution was lognormal.

Mathematically, it can be shown that the slope of the species–area curve, z, is 0.25 for a lognormal species–abundance curve. However, not all species–abundance curves are lognormal, and sometimes there are systematic deformations of the curve due to areal effects. Preston, for example, noted that small isolated islands had fewer species per unit area and a higher z parameter in their species–area curve than sample areas of the same size within continuous habitats on continents. He supposed that small isolated islands have fewer species because once a species became rare it was likely to become extinct, whereas in a continental habitat it can be sustained at low population levels by migration. Because small islands are more affected by the possibility of species extinction than large islands, the species–area slope is relatively steep compared with its continental counterpart.

Preston's observations are well illustrated in East Anglia, where Hooper (1970a) surveyed the ground flora of two groups of ancient woods which differed in their degree of isolation. The Cambridgeshire group had long been isolated in a landscape that has seen less than 2 per cent woodland cover over the last few hundred years. In contrast, the Rockingham Forest (Northants) is much more densely wooded and even today has more than 13 per cent cover. Hooper's data allowed Peterken (1981) to construct species–area curves for these two woodland groups (Figure 5.11). The Cambridgeshire woods have a z parameter of 0.5 as compared with only 0.2 for the Rockingham area, and the whole Cambridgeshire curve lies below that for Rockingham.

In one or two cases it has been possible to observe the recolonization of islands which, for one reason or another, have been completely denuded of their biota. The eruption of the Indonesian island of Krakatoa in 1883 is a classic example. What appears to happen to species numbers in such cases is that they rise fairly rapidly as recolonization takes place and then plateau

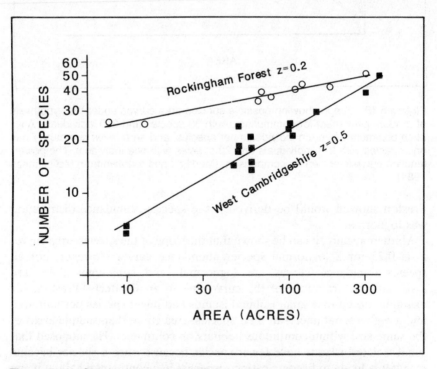

Figure 5.11 Species–area relationships for differing groups of East Anglian woods. The lower species richness of the West Cambridgeshire woods is a measure of their higher isolation. (Modified and redrawn from Peterken 1981)

to some equilibrium level. As new species arrive, others leave or become extinct through competition. Although species numbers may remain fairly level there is a turnover in the composition of species through migration and extinction. In the British Isles, detailed studies of bird species turnover, such as that described by Lack (1969) for the island of Skokholm, confirm a large number of immigrations and extinctions, and the existence of a species equilibrium level. On Skokholm, for example, during the period from 1928–1967 there were eight to nine regular breeding species with an average of one extinction and one immigration per annum.

MacArthur and Wilson (1967) brought together the three essential features of island biotas that have been described above, namely: species numbers increase with island area; species numbers remain steady but there is a continual turnover in species composition through immigration and extinction; isolation reduces the number of species. MacArthur and Wilson suggested that the number of species of a given taxon inhabiting an island represents an equilibrium between opposing rates of immigration and extinction. In fact, they define equilibrium as the state in which the rate of species extinction in a biota equals the rate of immigration of new species.

Suppose a new offshore island has been created, perhaps by changes in sea levels, and migration from the mainland species pool begins. At first the rate of immigration of new species to the island will be high because the probability that an immigrant species is new to the island is also high. With time, this probability will become less – hence the rate of immigration of new species declines and the number of species on the island increases. If all the mainland species capable of crossing water colonized the island, then the rate of immigration of new species would tend to zero but this of course does not happen. MacArthur and Wilson suggest that the immigration curve is concave because, on average, the more rapidly migrating species would be established first, causing a rapid initial drop in the overall immigration rate, while the late arrival of slow colonizers would drop the overall rate to an ever diminishing degree (see Figure 5.12). On the other hand, the rate of extinction will rise as more and more species colonize the island – since there are more species to become extinct and because average population size probably decreases due to competition for resources. MacArthur and Wilson reasoned that the extinction rate curve will probably be exponential in shape due to the compounding of diminishing population size and increasing competition between species. They argue that the actual shapes of the curves are not too important as long as they are monotonic (i.e. have only one turning point).

In Figure 5.12 there is one point where the immigration curve and the extinction curve cross. At this point the two rates are the same, and their level defines the turnover rate at equilibrium. The position of the curve intersection on the species axis defines the *equilibrium number of species*

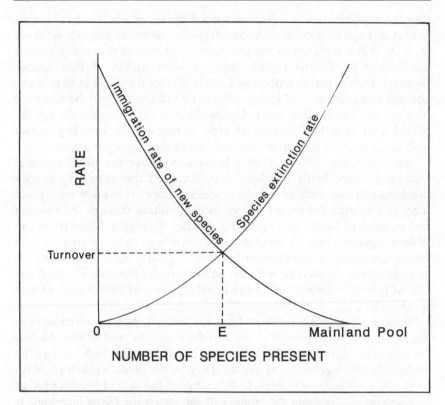

Figure 5.12 Equilibrium model of biota for a single island. The equilibrium species number (E) is reached at the intersection point between the immigration rate curve and the species extinction rate curve. (Modified and redrawn from MacArthur and Wilson 1963)

(E). Because the slopes of the immigration rate curve and the extinction rate curve vary with both island size and degree of isolation from the mainland species pool, diagrams showing equilibrium levels and turnover rates become more complex but provide some useful insights. In Figure 5.13 the equilibrium model predicts that turnover rates are highest on small, near islands, and lowest on large, far islands but species numbers at equilibrium are highest on large, near islands, and lowest on small, far islands.

The species–area relationship described above predicts the equilibrium numbers of species. Sometimes species–area data show an island to have too many species of a given taxon, or, perhaps, too few. It may be that the island is relatively young, or perhaps has undergone some tectonic movement which has affected its size. If the equilibrium model holds good on an island with too many species, then, given sufficient time, species numbers will relax to the new equilibrium level. Supersaturated large

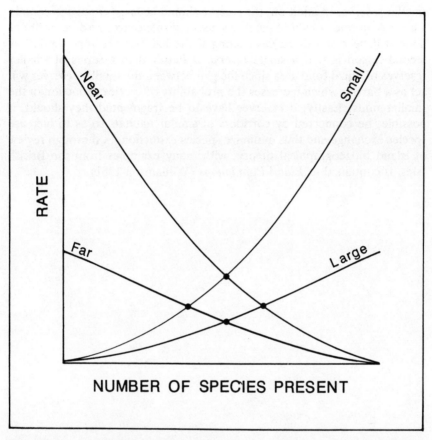

Figure 5.13 Equilibrium models of biotas of several islands of varying distances redrawn from the principal source area, and of varying size. (Modified and redrawn from MacArthur and Wilson 1963)

islands holding populations with low extinction rates will take longer to relax down to a new equilibrium than smaller islands.

MacArthur and Wilson's equilibrium theory of island biogeography is precisely that – a theory. It is a simple (some would say simplistic) model but has nevertheless acted as a catalyst for much research, and has some possible practical applications. In a world of increasing habitat destruction one of the most important ways of conserving species is the construction of nature reserves. These can be thought of as islands surrounded by landscapes given over to man's activities. The lessons of island bio-geography might be particularly useful in the design of such reserves where species must be protected from extinction if at all possible. The first lesson is that reserves should obviously be as large as possible, since a large reserve will support more species of a given taxon at equilibrium than a smaller reserve. In addition, the reserve should be large enough to contain a normal mosaic of habitat patches since disturbance to a reserve by fire or flood will be much more devastating if the habitats are very similar. A second lesson is that a single reserve is better than a series of smaller reserves of equal total area since the gap between the smaller reserves will act as a barrier which increases the probability of species extinction in the smaller units. Lastly, if reserves have to be fragmented they should, if possible, be connected by corridors of similar habitats so as to increase species exchange and thus minimize species extinction. A thorough review of island biogeographical theory, with many examples from the British Isles, is contained in *Island Populations* (Williamson 1981).

Chapter six

Limitations of the environment

In Chapter five we saw that, in some circumstances, species distributions are limited by intrinsic factors – such as competition and dispersal capabilities. This is only half the story, however, since we know that major environmental changes are frequently accompanied by the movement of biogeographical boundaries. As we shall see, the linkage between species distributions and environmental conditions is often very complex, and particularly so in the case of the British Isles.

Not only is it difficult to sort out what is a natural, as opposed to a man-influenced, distribution, but it is also often difficult to decide whether or not a particular distribution has had time to come into equilibrium with the environmental contraints, which themselves are often far from stable. For example, consider the possible impacts of global climatic warming as a result of the recent increase in greenhouse gases, and the destruction of the ozone layer. Other things being equal, even a 1 °C or 2 °C mean annual temperature increase over the next few decades could see a major northward expansion of many species in the British Isles.

Some words of caution

The survival of a species is possible only if environmental conditions do not exceed its range of tolerance, and if its minimum requirements are met. In biogeography, environmental tolerances and limits are often deduced from an examination of a species distribution map in conjunction with maps of environmental variables such as isotherms and isohyets. The simple geo-graphical technique of measuring spatial correlation raises many important issues of interpretation. Nobody would deny that environmental factors such as temperature and salinity have a profound affect on species distributions. However, distribution patterns alone rarely reveal processes, and the processes giving rise to spatial phenomena can be exceedingly complex.

Simple causal relationships imply the independence of environmental factors, which is rarely the case. In the British Isles, high mean annual air temperatures are significantly negatively correlated with mean annual rainfall, while in the shelf seas, mean surface water salinity in winter and mean surface water temperatures are positively correlated. Unless it is possible to pin down some physiological process to a particular environmental limit, the interpretation of spatial associations as being in some way causally connected must be treated with care.

Often we collect environmental data and produce some abstract statistic that has little meaning for a plant or animal. It is of no value in our understanding of the distribution of the common mole (*Talpa europaea*) to know the mean soil temperatures at the limits of its range. Moles, like other active mammals, experience temperatures continually and the concept of a statistical mean is not particularly applicable. But this also raises the point as to whether or not our measurements of environmental factors – such as rainfall and temperature – relate to the environment as experienced by the plant or animal. As Pigott (1975) points out, even the parts of a single plant may experience widely different temperatures on a warm sunny day. At noon, leaf temperatures on the sunny side of a hazel bush (*Corylus avellana*) may be as high as 25–30 °C, and those in the shade 15–20 °C, while the roots may be experiencing a temperature of about 10 °C. This range is as great as the difference in mean July temperatures between the Mediterranean and the north of Norway.

Climatic variations in relation to the longevity of plants further complicates the picture. For short-lived species, a run of bad summers might mean the failure to produce fertile seed, and extinction may result. On the other hand, long-lived trees, such as the small-leaved lime-tree (*Tilia cordata*), have an increased probability of producing viable seed in an exceptional year. In general, the fringes of most species distributions are non-breeding zones where adults can survive but not breed. For example, the spiny lobster or crawfish (*Palinurus vulgaris*) is common on southern and western coasts of the British Isles, but probably seldom breeds in these waters. *Palinurus* is a Lusitanian species, whose larvae are carried northward into the coastal waters of the British Isles where they metamorphose and complete their development.

For organisms in tidal habitats there is the added complication of deciding whether or not their distributions are controlled by aerial or marine environmental factors. The severe cold winter of 1962–3, for example, killed off the thick top-shell (or toothed) winkle (*Monodonta lineata*) at the eastern fringe of its range on the Dorset coast. When it was exposed at low tide, *Monodonta*, which usually clings to clean rock surfaces or seaweed, literally froze to death, its thick shell providing little protection against such extremes. In this example we have the notion of an extreme event influencing a distribution rather than some average. Similarly, the great storms in southern England in October 1987 literally clear-felled many small woods in Kent and Essex. Perhaps more magnitude-frequency studies and the recognition of the role of extremes are called for in the analysis of species distributions at the local, if not the national, scale.

A last point to bear in mind when investigating the influence of environmental factors on distributions is the quality of the data. Certainly, terrestrial distributions are reasonably well known, but there is still much to learn about marine distributions, which are intrinsically more difficult to determine. Of equal importance is the quantity of the environmental data. In the British Isles there is a paucity of climatological data for upland habitats, and isotherms, isohyets, and lapse rates are often little more than speculation and intelligent guesswork. In the marine environment, the problem is an order of magnitude more difficult in some respects because of its three-dimensional nature. This point was alluded to in Chapter four, where the concept of bathymetric sliding was described. As long ago as 1816, the French naturalist Lamouroux hypothesized that the organisms found at depth and on the ocean bottom were those found at the surface in the polar regions. Nearer home, bathymetric sliding is well illustrated by the levels of the seaweeds, bladder wrack (*Fucus vesiculosus*), and, immediately below it on the shore, serrated wrack (*Fucus serratus*). On the Isle of Man, the mid-tide level occurs at the junction of the two species, while on the Devonshire coast both species occur below this level.

Marine distributions

There are a number of useful sources of data on marine distributions. The *Provisional Atlas of the Marine Dinoflagellates of the British Isles* (Dodge 1981), the *Sea Area Atlas of the Marine Molluscs of Britain and Ireland* (Seaward 1982), the *Provisional Atlas of Marine Algae of Britain and Ireland* (Norton 1985), and maps from the *Continuous Plankton Recorder* (Colebrook *et al.* 1961) are all worth consulting, as is the *Journal of the Marine Biological Association*.

Temperature

Sea temperature is clearly the most important environmental factor influencing marine distribution patterns. This is not at all surprising, since marine organisms, with the exception of sea birds and mammals, are *poikilothermic*, their body temperature always being close to that of the surrounding water. This means that eurythermic species are most likely to be found in coastal waters where temperatures often fluctuate quite widely. It is also generally the case that marine organisms have a much more restricted range of tolerance during their early stages of development than as adults.

The American ecologist Hutchins postulated four types of marine distributional limits in relation to the latitudinal displacement of a species. These are:

(1) minimum temperature for survival;
(2) minimum temperature for repopulation;
(3) maximum temperature for repopulation;
(4) maximum temperature for survival.

Hutchins emphasized the point that both the survival of individuals and reproduction are necessary for sustained permanent establishment (repopulation). His four limits are shown diagrammatically in Figure 6.1. Distribution Type 1 would be expected where the poleward limits of a species range are controlled by its survival sensitivity to winter temperatures but the summer temperatures are warm enough to permit

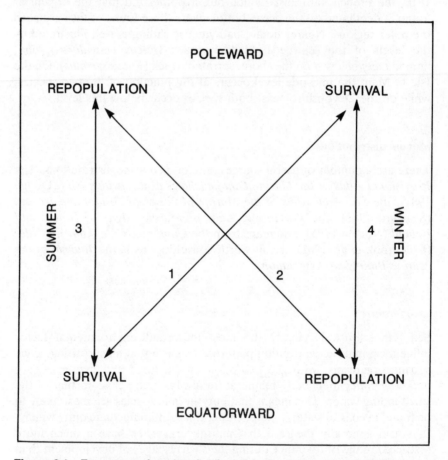

Figure 6.1 Four types of marine distribution in the northern hemisphere. (Modified and redrawn from Hutchins 1947)

reproduction and replacement. Toward the equator, the survival of the organism is limited by warm summer temperatures but with winter temperatures cool enough to permit reproduction and repopulation. A Type 2 distribution is regulated by temperatures required for the reproduction mechanism of the species without inhibiting its survival, i.e. summer temperatures poleward and winter temperatures equatorward. The poleward limits of Types 3 and 4 are the converse of Types 1 and 2 respectively (Hutchins 1947).

Implicit in Hutchins' system is the phenomenon of the longitudinal discontinuity of some species. These are stenothermal organisms which are found on the coast of Europe but are absent on the east coast of America because the temperature ranges are too great. One example is the pinkish hydroid (*Syncoryne eximia*) – a sea-fir – which is found on the lower shore in British waters, but is absent from American waters.

One of the most thorough investigations of longititudinal discontinuities and the effects of water temperature on species distribution is by Hoek (1982), who studied benthic marine algae in the north Atlantic Ocean. Hoek was able to formulate rigorous hypotheses concerning the temperature responses which determine distribution boundaries because he had observations both on the western and eastern Atlantic coasts. Annual temperature fluctuations along the west Atlantic seaboard are much more pronounced than on the east Atlantic coast. With species limited to only one side of the ocean, boundaries can, with equal probability, be determined by summer or winter temperatures. Furthermore there is a large literature relating to the experimental verification of temperature regulation of benthic marine algal life histories. Hoek suggested six types of critical temperature limiting essential events in algal life histories which might determine a distribution boundary:

northern lethal boundary: the lowest winter temperature a species can survive for a period of 2–4 weeks;

southern lethal boundary: corresponds to the highest summer temperature a species can survive during a period of 2–4 weeks;

northern growth boundary: the lowest summer temperature which over a period of several months permits sufficient growth;

southern growth boundary: the highest winter temperature which over a period of several months permits sufficient growth;

northern reproductive boundary: the lowest summer temperature which over a period of several months permits sufficient reproduction;

southern reproductive boundary: the highest winter temperature which over a period of several months permits sufficient reproduction.

Hoek's boundaries are seen to be a refinement of Hutchins' scheme.

As we learnt in Chapter two, sea temperatures around the British Isles vary considerably, and this is reflected in major zonation of many taxa into

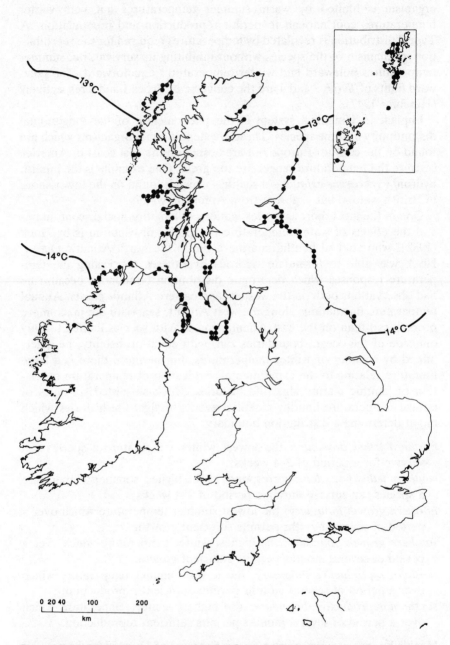

Figure 6.2 Distribution of the red alga (*Odonthalia dentata*), a common species in Arctic waters. Isotherms are mean summer surface water temperatures. (Modified and redrawn from Norton 1985)

northern or southern species. The dividing line is about the 10 °C mean annual surface isotherm – although this is only a rough guideline. What is perhaps more interesting is that the waters around the British Isles are essentially the last port of call for northern species reaching down from the Arctic and for southern species reaching up from North Africa and the Mediterranean.

Several northern seaweeds have well-defined southern limits on the coasts around the British Isles. This is clearly seen in the distribution of *Odonthalia dentata* which is one of the most distinctive sub-tidal red algae, and is often cast up by waves. This alga is common in Arctic waters, but reaches no further south than the Isle of Man in the British Isles (Figure 6.2).

Among the fish species more common in cold northern waters north of the 10 °C mean annual isotherm are the cod (*Gadus morhua*), the haddock (*Melanogrammus aeglefinus*), and the ling (*Molva molva*). In the warmer waters of the English Channel, the Western Approaches, and southern North Sea Basin, fish species such as pollack (*Pollachius pollachius*), Dover sole (*Solea solea*), and pilchard (*Sardina pilchardus*) are more common. In the last 25 years or so, mean annual water temperatures in the seas around the British Isles have fallen slightly as compared with the previous couple of decades, and the distribution of many marine species has shifted slightly southward as a result. In the English Channel, cod, ling, and herring have become more numerous, whereas warm-water species such as pilchards and hake (*Merluccius merluccius*) have become less common (Tait 1981).

Many southern taxa show a marked western bias in their distributions and are absent from the North Sea (Table 6.1). This probably reflects the fact that the isotherms for their northern lethal boundaries are several degrees higher on western coasts than at the same latitude on the east (see Figure 6.3).

Precisely what controls the distribution limits of species can sometimes be determined through transplant experiments. One species which has

Table 6.1 The limits of some southern species around the British Isles

Species	Northern limit	Eastern limit	Comments
Balanus perforatus (an acorn barnacle)	Pembroke	Isle of Wight	Absent Ireland
Didemnum maculosum (a sea squirt)	N. W. Scotland	?	Absent North Sea
Anemone sulcata (snakelocks anemone)	N. W. Scotland	Isle of Wight	Absent North Sea
Bifurcaria bifurcata (a brown seaweed)	Sligo	Devon	Absent North Sea (see – Figure 6.3)

Source: Lewis (1976)

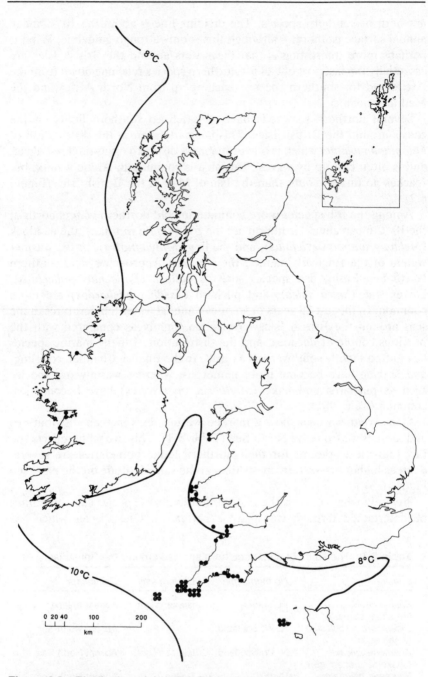

Figure 6.3 Distribution of the alga *Bifurcaria bifurcata* – southern species of sheltered shores. Isotherms are mean winter surface water temperatures. (Modified and redrawn from Norton 1985)

been studied in this manner is the large brown seaweed, dabberlocks (*Alaria esculenta*) (Sundene 1962). This is a northern species, found on exposed coasts around the British Isles, attached to rocks just below low-tide level. Its southern lethal boundary is closely related to the 16 °C summer isotherm (see Figure 6.4). The absence of *Alaria* from the southern North Sea Basin could also be due to a change in environment: from rocky coasts exposed to the Atlantic swell to more protected shingle and sand. In order to distinguish between these possibilities, mature samples of *Alaria* were transferred one autumn to a sheltered rocky site in the Oslofjord in eastern Norway, where the species does not normally grow. In the following spring most of the plants grew rapidly but by July/ August water temperatures in the fjord had reached 17–18 °C, the plants

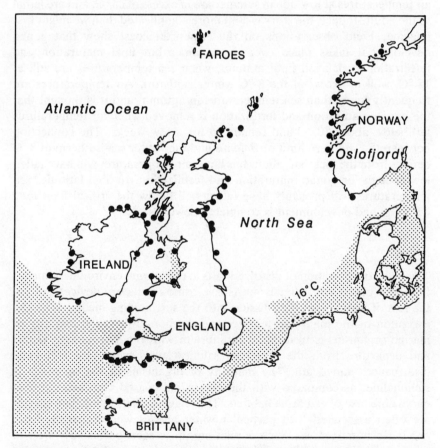

Figure 6.4 Distribution of the alga *Alaria esculenta*. Shaded areas have mean summer surface water temperatures in excess of 16 °C. (Redrawn and modified from Dring 1982)

had stopped growing, and their thalli were breaking up. By September all the transplanted *Alaria* had died. Spring water temperatures were evidently high enough for asexual reproduction since zoospores had been produced and had germinated into minute sporophytes by April, but by mid-summer these suffered the same fate as the adult transplants. Culture experiments with young *Alaria* sporophytes showed rapid growth at 14 °C and no growth at 17 °C.

Balanus balanoides is an arctic species of barnacle found around the coasts of the British Isles but absent from the Scilly Isles and the tip of Cornwall. Its southern limit corresponds to the 8 °C winter surface-water isotherm. As was mentioned previously, a single boundary has two interpretations. In the case of *Balanus*, it might be that water temperatures south of this isotherm fail to fall low enough for reproduction, or perhaps air temperatures at low tide in summer are in excess of 25 °C, and are lethal to the adult. In fact, the story is a bit more complicated than we might first imagine. Field observations on the American coast show that gonad development takes place over the summer but final maturation and fertilization is delayed until autumn, when sea temperatures are still at 15 °C, well in excess of the 8 °C winter isotherm. Air temperatures are frequently lower than sea temperatures in autumn, and it is thought that the bar to maturation and fertilization is removed when air temperatures fall below about 10 °C and remain so for a few weeks. The connection between the southern limit of *Balanus* and the winter sea isotherm of 8 °C derives from the fact that, about this limit, air temperatures will have fallen low enough for gonad maturation and fertilization. In high latitudes sea temperatures will probably have fallen well below the critical limit long before gonad development is complete (Lewis 1976).

Salinity

The high concentration of dissolved salts in the marine environment means that both plants and animals must have osmo-regulatory devices to limit the loss of water from their tissues into the surrounding media. With the exception of the higher vertebrates and the bony fishes, the majority of marine organisms is in osmotic equilibrium with the surrounding water, and departure from the preferred salinity levels will cause metabolic disturbances and death. The majority of organisms in the open sea is stenohaline, as compared with the euryhaline coastal taxa and extreme euryhaline taxa of estuarine habitats. This suggests that stenohaline species are often associated with particular water masses whose characteristics are fairly homogenous. Such a relationship is seen particularly well in plankton, where so-called *indicator species* demarcate the major water bodies around the British Isles.

Plankton distributions can be examined by means of a T–S graph (Bary

1963). The two axes of a T–S graph are water temperature and salinity. Within the graph, the boundaries of major water masses can be defined and the characteristics of the waters containing a plankton species plotted. Figure 6.5 is a typical T–S graph for the north-east Atlantic and North Sea during late summer. Water samples containing three species of plankton have been plotted, and it can be observed that there is a clear separation of these plankton in terms of major water mass characteristics. *Calanus minor* is restricted to warmer oceanic water, *Pleuromamma robusta* to colder oceanic water, and *Sagitta elegans* only occurs in mixed shelf-water.

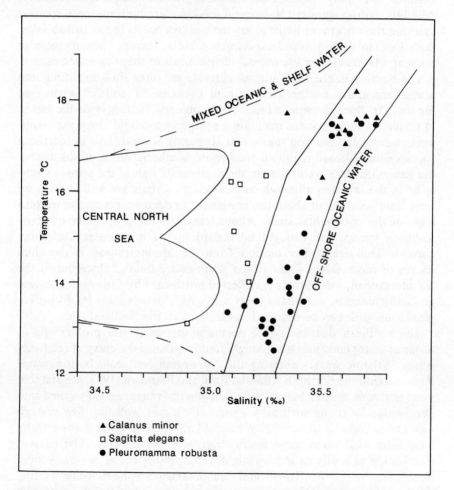

Figure 6.5 Temperature–salinity plankton diagram for the north-east Atlantic and central North Sea, August–September, 1957. (Modified and redrawn from Bary 1963)

Sagitta elegans is, in fact, one of the best-known indicators of the presence of Atlantic coastal (neritic) water in the North Sea.

Tides and currents

Many marine species produce propagules, such as spores (plants) or free-swimming larvae (animals), which are dispersed by water currents, tidal streams, and residual drifts. According to Earll and Farnham (1983) the distributions of many marine animals in the shallow sub-littoral zone are strongly influenced by the pattern of surface- and deep-water currents. For example, the spiny starfish (*Marthasterias glacialis*) and a sea cucumber, called the cotton spinner (*Holothuria forskali*), are both southern species, reaching their northern limits along the western coasts of the British Isles. Both Lusitanian and mixed Lusitanian/Atlantic current patterns seem to account adequately for the overall distributions of these two species.

The North Atlantic Drift surface-currents are often shown splitting into a southern and a northern component between 40° and 55° north (see Figure 2.3). But this pattern raises two problems. First, how do the larvae of sessile southern species maintain a northerly foothold along the south-west coast of Ireland and south-west England? Second, how do northern species establish and maintain their more southerly distributions against the general surface-water drift to the north-east? Part of the answer seems to lie in the fact that although surface-water currents are well known, we have little good detail about the movement of deep waters on the western edge of the continental shelf. Almost certainly the propagules of many southern species are brought northward in the deep-water Lusitanian current. This deep-water outflow from the Mediterranean is the chief source of more saline water in the north-east Atlantic. Once out of the Mediterranean, this current is deflected northward by Coriolis force, and gradually ascends, eventually to overlie cold, Arctic water. Its distinctive planktonic drift can be detected as far north as the Shetland Isles.

The southern distribution of northern species is still problematic in terms of water currents. It is thought that to balance the entry of relatively warm, Atlantic waters entering the Norwegian Sea, cold bottom-water flows out into the northern Atlantic. Earll and Farnham (1983) suggest that most northern species have a wide bathymetric range, and if washed into deep-water locations such as sea-lochs, they may well find low enough water temperatures to survive. We should also bear in mind that we really have little idea when these distributions were established. The present distribution of northern species might well be partly relict, having become established when the British and Scandinavian ice-sheets were wasting northward some 10,000 years ago.

The successful dispersal of the free-swimming stage of many intertidal or neritic animals also depends on its duration. So for example, in spite of the fact that currents should be able to transport the larvae of the ormer (*Haliotis tuberculata*) – from their northernmost outpost in the Channel Isles to the Devon and Cornish coasts – the species is absent. Evidently, the free-swimming stage of this southern species of limpet has a duration of only about 30 hours which is not long enough for it to drift to the English coast.

Substrate

The nature of the coast and the sea floor can be important factors in the distribution of sessile neritic species and some deeper water benthic species. Animals such as mussels and limpets that require hard substrates are not found along sandy coasts; likewise, animals that need to burrow in mud and sand are usually absent along rocky stretches of coast. The distribution of several species of demersal fish is also closely associated with the nature of the seabed. In the North Sea the sand-eel (*Ammodytes marinus*) collects in dense schools over the tops of sand ridges, and on the edges of sandbanks such as the Dogger Bank (the generic name *Ammodytes* means 'sand digger'). Sand-eels remain buried in sand at night, and also during winter when light intensities are low. The Dover sole is particularly abundant in the southern North Sea and Irish Sea – in areas of muddy or fine sand where its main food, the ragworm (*Nereis* spp.), is abundant.

In their study of the distribution of intertidal organisms along the English Channel, Crisp and Southwood (1958) suggest that, in addition to variations in temperature and salinity, there is an important contrast in the substrate between the western and eastern basins (broadly divided along a line between the Isle of Wight and the Cotentin Peninsula). In the eastern basin hard rock outcrops are infrequent, and the coast is dominated by shingle beaches – whose material is a very effective rock-scouring medium. As far as we know, all species found in the eastern basin are also found in the western basin, but many western species are absent from the eastern basin of the Channel. For example, several species of algae – such as *Bifurcaria bifurcata, Himanthalia elongata*, and *Saccorhiza polyschides* – are absent eastward of the Isle of Wight. Crisp and Southwood suggest that this is due to lack of sufficient rocky substrate, damage due to scour, and opacity of the water.

Terrestrial distributions

Undoubtedly the best sources of data for studying the distributions of terrestrial taxa are the various atlases published by the Biological Records Centre, often in conjunction with a scientific society (see Chapter one). Of particular value are the transparent environmental overlays, which have

141

been produced by the Biological Records Centre at the same scale as the maps in the Atlas series (Institute of Terrestrial Ecology 1978). *The Atlas of the British Flora* (Perring and Walters 1962) and the *Critical Supplement* (Perring 1968b) remain essential sources for data on vascular plant distributions.

Temperature

This factor probably influences the distribution of terrestrial taxa more than any other and has been particularly studied in plants (Conolly and Dahl 1970; Bannister 1976; Carter and Prince 1985). Temperature affects all stages of plant development: from the production and dispersal of propagules, to their germination, through to photosynthesis in the mature plant.

The effect of temperature on seed production in the small-leaved lime tree (*Tilia cordata*) has been thoroughly investigated by Pigott and Huntley (1981). The northern limit of the natural distribution of Tilia is in the English Lake District, but in north-west England there is no regeneration from seed, and even when seeds are produced they are infertile. Pigott and Huntley discovered that, under the present climate in the Lake District, the pollen-tube fails to grow down the style of the female flower sufficiently to fertilize the ovary and that the rate of growth of the pollen-tube is dependent on the temperature during the flowering season. Frequent seed production in *Tilia cordata* rarely occurs where the average mean daily August maximum temperature is less than 20 °C. *Tilia cordata* must therefore have spread into north-west England when temperatures were about 2 °C higher than the 18 °C or so presently found in the region (see Figure 6.6)

In point of fact, *Tilia* could only have spread northward from the Continent during the post-glacial period when such warm conditions prevailed. Pigott and Huntley's findings illustrate the important point that although there is a causal relationship between climate and the distribution of *Tilia cordata* in northern England it is no longer direct. Pigott (1975) also studied the stemless thistle (*Cirsium acaulon*). He found that its south-eastern distribution was not only due to a failure to produce seed in more northerly locations but also because in these localities lower evaporation rates – due to lower temperatures – encouraged the growth of the fungus *Botrytis cinerea* just beneath the flower heads, causing them to fall off.

Even if seeds are successfully produced they may require temperature pretreatment before their dormancy is broken and germination can take place. The cloudberry (*Rubus chamaemorus*) is a herb of the high moors and blanket bogs in northern England and Scotland, and its seeds require at least 5 months' low-temperature stratification at 4–5 °C to break dormancy (Bannister 1976). This low temperature requirement suggests

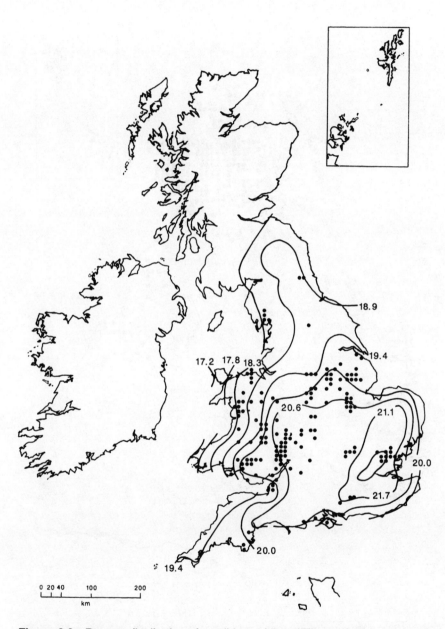

Figure 6.6 Present distribution of small-leaved lime (*Tilia cordata*) in relation to isotherms of average means of daily maximum air temperatures (reduced to sea level) during August for the period 1901 to 1930. (Modified and redrawn from Pigott and Huntley 1983)

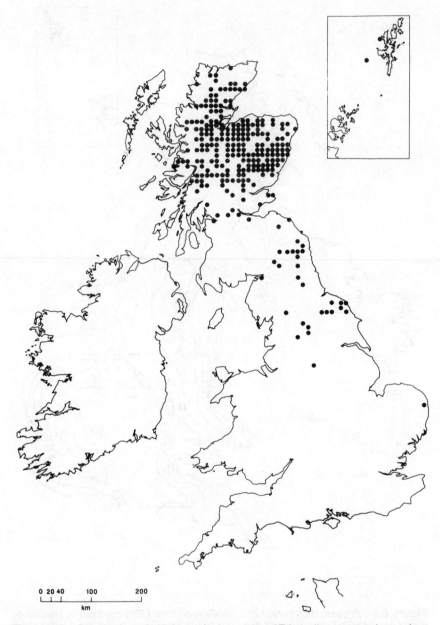

0 20 40   100   200
km

Figure 6.7 Distribution of chickweed wintergreen (*Trientalis europaea*) – a plant limited to cold upland areas. Low temperature stratification increases its germination rate (Modified and redrawn from Perring and Walters 1962)

that even if its seeds were dispersed to lower altitudes, they would not germinate because of the high temperatures. In some instances, low temperature stratification also increases the germination rate, as in the case of chickweed wintergreen (*Trientalis europaea*), which is locally common in pine-woods and among moss in grassy and boggy places in Scotland and the northern Pennines (see Figure 6.7).

Salisbury (1939) suggested that annual plants can be grouped into two categories, namely the *summer annuals* and the *winter annuals*, according to the time of germination of the seeds. Summer annuals, such as fluellen (*Kickxia spuria*) – found in southern England, Wales, and southern Ireland – pass the winter season in the resistant seed phase, and do not germinate until the advent of warm conditions in spring. This commonly results in a relatively late production of flowers, and seeds fail to ripen in northern regions and hence the northward extension of these species is restricted. In general there is a good correspondence between the distribution of some summer annuals and the mean maximum temperature for the month in which their seeds usually ripen, namely September. Winter annuals, such as striated catchfly (*Silene conica*), shed their seeds early in summer, and these germinate in the autumn to produce a hardy rosette stage. *Silene conica* can withstand fairly continental climatic regimes and is widely distributed in central Europe.

The effects of temperature on the balance between photosynthesis and respiration may well be important in determining a plant's climatic limits. Both low and high temperatures may produce situations where the apparent photosynthesis is zero, because the output of carbon dioxide by respiration and its uptake by photosynthesis are identical. Such situations are called *compensation points*. Somewhere between the low and high temperature compensation points net assimilation occurs at an optimum temperature which is usually raised as light intensity increases. Both altitudinal limits and latitudinal limits may thus, in part, be decided by the delicate balance between photosynthesis and respiration. Failure to accumulate food reserves may cause a lack of competitive vigour, failure to flower and set seed, or even lay the plant open to attack by pathogens.

The balance between respiration and photosynthesis is probably an important factor in determining the altitudinal limits of trees such as the Scots pine (*Pinus sylvestris*). This species has an altitudinal limit of about 1,000 m in western Norway. In eastern Scotland, the limit is down to about 700 m, and on the west coast of Scotland it is lower still at about 100 m. In the oceanic climate of western Scotland photosynthesis is limited at higher altitudes by low air temperatures, increased cloudiness in summer, and by short day length and low light intensities in winter. But the relatively high winter temperatures induce high respiration rates leading to a depletion of carbohydrate reserves and loss of vigour (Bannister 1976). At lower altitudes it is less cloudy and conditions are more favourable for photosynthesis. Under a more continental climatic regime, the warmer and less cloudy summers allow more photosynthesis, and the colder winters minimize respiratory losses, thus allowing trees to grow at a much higher altitude.

Evergreen Arctic-alpine species, such as the purple saxifrage (*Saxifraga oppositifolia*), and the mountain avens (*Dryas octopetala*), have the ability to photosynthesize at low light intensities and temperatures, and are usually found above the treeline away from the competition of larger species. However, in western Scotland and on the karstic tableland of the Burren in western Ireland, arctic-alpine species may be found close to sea level, where cool summers and lack of competition allow such species to extend their range.

The effect of temperature on vegetative growth, and on subsequent competitive vigour and distribution, has frequently been reported – for example, Pigott (1975), Woodward (1976), and Woodward and Pigott (1976). The rose-root (*Sedum rosea*) is an arctic-alpine species, common in high altitude rock crevices northward from south Wales and Yorkshire, and on sea-cliffs in western Scotland and Ireland. In the Lake District and the Pennines it is not found below 450 m OD. In contrast, the orpine (*Sedum telephium*) is found in woods and hedge-banks over the greater part of the British Isles, and reaches an altitudinal limit in the Lake District of 460 m OD at Kentmere. Pigott and Woodward carried out a series of pot-transplant experiments to see how these two species grew at different altitudes. They found that both grew healthily over a wide range of altitudes. The growth rate of the lowland species declined linearly with altitude, while that of the upland species was scarcely affected. The vigorous growth of *Sedum telephium* at low altitudes enables it to compete with other species, but with increasing altitude its competitive potential declines, until it is less than that of *Sedum rosea*. As Pigott points out, a study of *Sedum rosea* alone would have provided no indication of the conditions controlling its distribution.

The interplay of summer and winter conditions on plant distributions was studied by Iversen (1947), who examined the factors controlling the distribution of mistletoe (*Viscum album*), ivy (*Hedera helix*), and holly (*Ilex aquifolium*) in north-west Europe. Iversen collected temperature data from meteorological stations known to be near the geographical limits of these species. The mean temperatures of the warmest and coldest months were recorded, as was whether or not a species was located within 20 km of the station. If a species was present, it was also noted if it bore fruit or was sterile. These data allowed Iversen to construct a scatter plot, on which a line could be drawn between stations having and not having the species present. This line, or envelope, Iversen called the *thermosphere*. An Iversen plot for mistletoe is shown in Figure 6.8, together with its distribution in the British Isles. Where summer temperatures are high

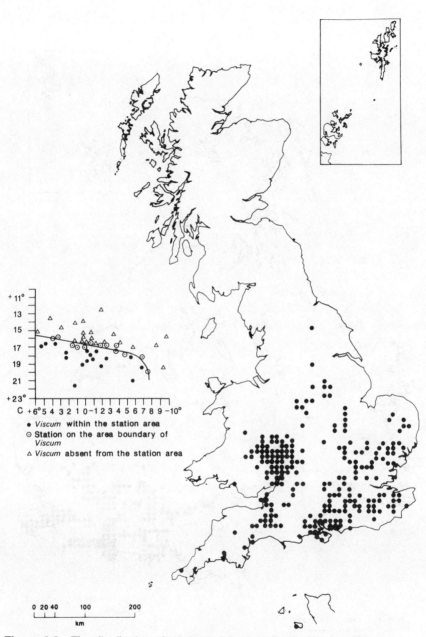

Figure 6.8 The distribution of mistletoe (*Viscum album*). The inset diagram shows the thermosphere as defined by Iversen (horizontal axis – mean temperature of coldest month; vertical axis – mean temperature of warmest month). Mistletoe is absent redrawn from Ireland. (Based on Perring and Walters 1962; Iversen 1944)

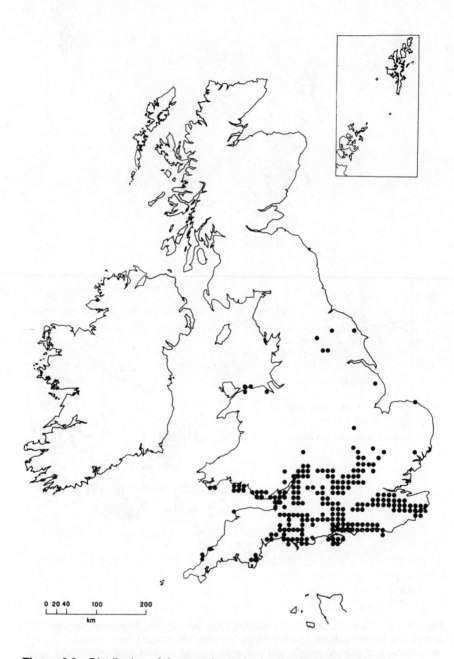

Figure 6.9 Distribution of the round-mouthed snail (*Pomatias elegans*) – a southern species of calcareous soils. (Redrawn from Kerney 1976)

enough, mistletoe can tolerate winter temperatures as low as $-8\,°C$, but in regions of lower summer temperatures its cold tolerance decreases. This explains the location of mistletoe in the south-east of the British Isles, where continentality is greatest.

Homoeothermic animals in low air-temperatures require only a little more food to counteract the increased loss of heat, but for poikilothermic animals, and many plants, winter cold creates problems of survival. Often they avoid such stress by dormancy. In insects, this quiescent stage, or diapause, is one in which they are *just* metabolizing but all growth is arrested. Any of the life-history stages – egg, larva, pupa, or adult can undergo diapause. Most beetles, for example, are dormant over the winter period, and young adults come out of hibernation during the spring when the temperature rises above their activity threshold. Similarly, slugs and snails make soil burrows in the autumn where they hibernate during the winter months. This activity suggests that the distribution of many invertebrates is governed by the annual minimum (probably soil) temperature which, in some cases, is very low. The asparagus beetle (*Crioceris asparagi*) for example, can withstand temperatures as low as $-23\,°C$. The distribution of molluscs such as the round-mouthed snail (*Pomatias elegans*) is also probably controlled by temperature factor. The distribution of this snail is essentially limited to calcareous soils in southern England and Wales. It is noticeably absent from the more continental climates of East Anglia where winter temperatures are low and frost frequent (Figure 6.9).

In high uplands low temperatures result in the disturbance of the often thin skeletal soils by frost, and this factor, combined with the severe exposure, probably explains the paucity of animal species at high altitude. Some moths, such as the northern eggar (*Lasiocampa quercus callunae*) and butterflies, such as the small mountain ringlet (*Erebia epiphron*) are only found at high altitude in Britain and are associated with damp *Nardus* grassland. This fact alone probably stops them spreading to lower sites. A common adaptation of montane insects is either a brown or black colouration, which lowers the insect's albedo and allows it to absorb the maximum amount of radiation during the short summers.

During the unfavourable season the perennating buds which (except in annual plants) give rise to new shoots, and thus carry on the life of the plant, are at risk from frost and loss of water. The Danish botanist, Raunkiaer, suggested that it was possible to classify plants according to the position of the perennating organ with regard to ground level and that this position was correlated with the regional climates. Raunkiaer's classification of so-called *life forms* is illustrated in Figure 6.10. The chief categories are as follows:

phanerophytes – buds and shoots are borne well above ground (broadly trees and shrubs). Subdivided into: *mega- and mesophanerophytes*, taller

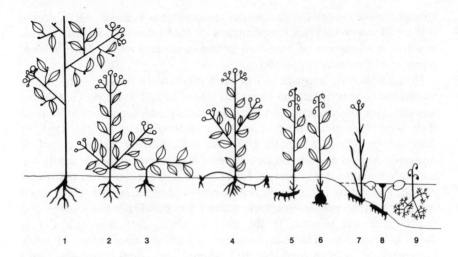

Figure 6.10 Main type of life form based on Raunkiaer's classification: 1 phanerophytes; 2–3 chamaephytes; 4 hemicryptophytes; 5–6 cryptophytes; 7 helophytes; 8–9 hydrophytes.

than 8m; *microphanerophytes* 2–8m tall; *nanophanerophytes*, 0.25–2m tall;

chamaephytes – buds raised into air but not more than 0.25m above ground (undershrubs or herbs);

hemicryptophytes – buds formed in the surface soil;

geophytes – buds formed in soil on corm, bulb, or tuber (also called cryptophytes);

helophytes – marsh plants with buds in soil or mud below water level;

hydrophytes – buds situated under water (water plants);

therophytes – annual plants, but includes 'winter annuals' whose seeds germinate in the autumn and seedlings survive winter as a short stem with a rosette of leaves.

Raunkiaer was particularly interested in the percentage distribution of the different life-form classes in the total flora of a region. He gave the term '*biological spectrum*' to such a distribution list and showed quite clearly that it was related to the climate. Table 6.2 shows the percentage distribution of each life-form class in an altitudinal sequence at Glen Clova in the south-east Cairngorms, Scotland.

The biological spectrum at Glen Clova is dominated by chamaephytes and hemicryptophytes. These life forms are successful here because the winter snows protect their perennating buds from wind-chill. At Clova, the highest chamaephyte percentage is not reached until +1,000 m. By way of contrast, in the more continental climate of the Swiss Alps, the 27 per cent

Table 6.2 The variation of Raunkiaer's biological spectrum with altitude – Glen Clova, north-east Scotland.

Altitude	No. of species	Life-form class (per cent)							
(m)		MM	M	N	CH	H	G	HH	TH
1,000+	11				27	64	9		
900–1,000	44			2	25	52	14		7
800–900	72			3	22	60	11		4
700–800	170		1	5	11	67	11	1	4
600–700	206	2	2	4	15	62	9	2	5
500–600	182	3	3	4	13	63	8	1	4
400–500	193	3	3	4	11	66	8	1	4
300–400	211	3	3	4	10	65	8	2	5
–300	304	3	2	4	7	59	7	5	13

Source: Raunkiaer (1934). Mega and mesophanerophytes (MM); microphanerophytes (M); nanophanerophyte (N); chamaephytes (CH); hemicryptophytes (H); geophytes (G); hydrophytes and helophytes (HH); therophytes (TH).

chamaephyte level is not reached until 2,500–2,700 m, and in the Faeroe Isles, some 500 km north of Scotland, it is reached by 700 m. These three results illustrate very clearly the influence of general climatic trends on biological spectra. Raunkiaer thought that a knowledge of such spectra was fundamental for the definition of plant climatic boundaries or *biochores*. At the macro-climatic scale, he was able to use the rise in the chamaephyte percentage to distinguish three biochores along the following latitudinal transect:

cold temperate climate	Scilly Isles (CH = 3.5%)
(hemicryptophyte zone)	Shetlands (CH = 7.0%)

10% chamaephyte biochore = 10 °C June isotherm

boreal climate	Iceland (CH = 13.0%)
(hemicryptophyte–chamaephyte zone)	

20% chamaephyte biochore = 4.4 °C June isotherm

arctic climate	East Greenland (CH = 25.0%)
(chamaephyte zone)	(75°N)

30% chamaephyte biochore

arctic nival climate	Jan Mayen Island (CH = 32.0%)
(chamaephyte zone)	

At the regional scale, the relationship between climate and the biological spectrum is less precise, and more dependent on factors such as geology and soils. Shimwell (1971) has investigated the available data thoroughly, and suggests a threefold division of the British Isles:

1 chamaephyte zone – highland Britain and north-east Ireland;
2 hemicryptophyte zone – lowland Britain and south-west Ireland;
3 therophyte zone – East Anglia, especially around Breckland.

Precipitation

The distribution of precipitation in the British Isles is an important secondary factor influencing the distribution of many taxa, although it is sometimes difficult to isolate its gross effect from local influences – such as soil texture and slope. For many invertebrates unable to tolerate much water loss there is a requirement for a highly humid environment, which is controlled by the precipitation/evaporation ratio. Humidity probably accounts for the distribution of a number of species of land snail, such as *Zenobiella subrufescens* and *Pyramidula rupestris*, which frequent the north and west of the British Isles but are absent from the drier parts of East Anglia and the south-east of England. On the other hand, several snail species prefer xerophytic habitats, such as the calcareous grasslands of southern and eastern England. Examples are *Pupilla muscorum* and *Vallonia costata*, both of which have relatively thick shells which may help to impede water loss.

Ratcliffe (1968) examined the relationship between climatic variables and liverworts and mosses found on the Atlantic fringes of the British Isles. The majority of species require a wet climate, and Ratcliffe indicates that the distribution of rainfall may be more important than the actual total because a high number of rain days militates against any prolonged, damaging drought. Although many parts of the Lake District and North Wales have an annual rainfall in excess of 2,000 mm, they do not have the same species richness as parts of western Ireland and the western Highlands of Scotland, where total rainfall may be less than 2,000 mm per annum but where there is a more even distribution throughout the year. Figure 6.11 shows the distribution of the liverwort *Adelanthus decipiens*, which requires a highly oceanic climate. Its failure to colonize the very north-west of Scotland and the wetter parts of the Pennines is probably due to its dislike of long-lying snow.

Some plants, such as the spanish catchfly (*Silene otites*), positively avoid high precipitation levels. This local species is confined to the dry, sandy soils of the Breckland, East Anglia, which is probably the most arid environment in the British Isles. Annual rainfall in the Breckland is less than 600 mm, and the summer soil moisture deficit is in excess of 100 mm.

Soils

Edaphic influences on distributions are most clearly seen in those species that have a requirement for calcium, which is freely available in soils having a pH in excess of 7. Over chalk and limestone, rendzinas and calcareous brown earths typically carry a calcicolous flora, and a rich mollusc fauna. At the coast, calcium is also available through the comminution of shelly materials, which are easily wind-blown. On many wet

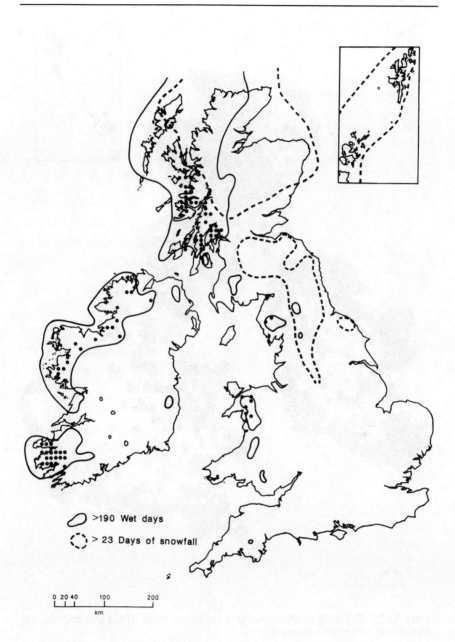

>190 Wet days

> 23 Days of snowfall

0 20 40   100   200
km

Figure 6.11 Distribution of *Adelanthus decipiens* in relation to indices of limiting humidity and temperature. This liverwort species is confined to areas where there is high atmospheric moisture and freedom redrawn from low winter temperatures. (Modified and redrawn from Ratcliffe 1968)

(a)

0 10 20   100   200

km

Figure 6.12 Generalized distributions of (a) bell heather (*Erica cinerea*) and (b) kidney-vetch (*Anthyllis vulneraria*)

moorlands in the British Isles the only mollusc present is the large black slug *Arion ater*, a species practically without a shell. Molluscs in general require calcium to build their shells, and their distribution is limited by its availability.

(b)

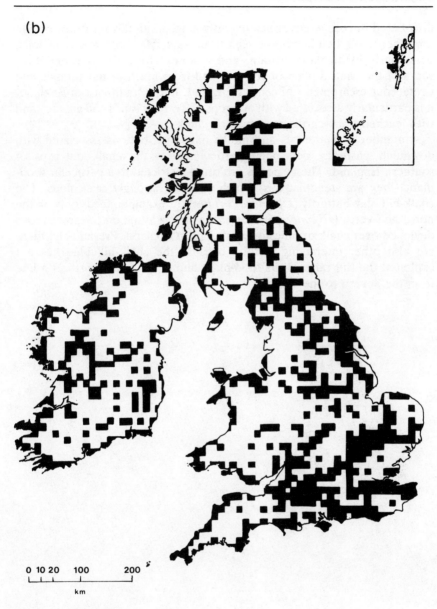

0 10 20   100   200

km

Figure 6.12 continued

Figure 6.12 shows the distribution of the calcicole, kidney-vetch (*Anthyllis vulneraria*) and the calcifuge, bell heather (*Erica cinerea*) both of which are widespread in the British Isles. The kidney-vetch shows a clear preference for soils over chalk or limestone, but also is widely

155

distributed in coastal environments on young, wind-blown substrates. On the other hand, bell heather is seen to be associated with wet, acid soils, particularly in highland Britain, and is absent from the Carboniferous rocks of the central plain of Ireland, and from much of the Jurassic and Cretaceous escarpments of central England. In the south-east of England, it is particularly associated with outcrops of Tertiary sands and gravels, and with leached brick-earths preserved on chalk.

A number of butterflies and moths have distributions associated with downland pastures, developed particularly on the chalk outcrops of southern England. These species are often restricted to a particular food-plant (they are stenophagous) which is, itself, strongly calcicolous. The chalk-hill blue butterfly (*Lysandra coridon*) for example, feeds only on the horseshoe-vetch (*Hippocrepis comosa*), which is found on calcareous soils devloped over chalk or limestone outcrops in England. Presumably, there are also other interacting factors, because the chalk-hill blue has not exploited the full range of its food-plant and is not found north of a line from the Severn to the Wash.

Chapter seven

Geographical relationships

In the last two chapters we have seen that a species distribution is the product of intrinsic and extrinsic factors both of which govern its possible geography. In this chapter we shall see that groups of species do, in fact, have remarkably similar geographies. By this, we mean that their distributions are broadly the same, both within the British Isles and on the surrounding continents. The recognition of these different geographical groupings, and how these came about, both in time and space, is arguably at the very heart of biogeographical enquiry.

Geographical ranges on land

More than a century ago, both Watson (1847) and Forbes (1846) independently produced a classification of the geographical ranges of taxa into what we would now call phytogeographical and zoogeographical regions, although their purpose for doing so was very different.

In the *Cybele Britannica*, Watson grouped the British flora into seven types:

1 *British type* – widely occurring in England and Scotland (alder, hazel, meadow buttercup);
2 *English type* – a predominantly English distribution (small-leaved lime, water violet);
3 *Scottish type* – a predominantly Scottish distribution (globe flower, Scots pine, crowberry);
4 *Highland type* – mountain plants of Wales, Scotland, England (dwarf willow, mossy cyphel);
5 *Germanic type* – plants characteristic of eastern England (pasque flower, grape-hyacinth);
6 *Atlantic type* – plants of the west and south-west Britain (pennywort, Dorset heath, English stonecrop);
7 *Local or doubtful type*.

Watson was a great collector of facts, and his classification is best thought of as an original attempt at what geographers would now call spatial differentiation, based mainly on a climatic zonation. Apparently, Watson had little interest in the geographical origins of the flora, and this led to the great disadvantage of his classification, namely, that it does not take into account the nature of a species range beyond Britain. When the European range of a species is examined, it is often apparent that the British portion of its geographical area connects characteristically with a particular type of continental distribution. The insensitivity of Watson's grouping is seen clearly if we examine the wider distribution of plants in his British type. Species such as the bluebell (*Endymion nonscriptus*), the cross-leaved heath (*Erica tetralix*), and the foxglove (*Digitalis purpurea*) occur widely in the British Isles where suitable habitats exist, and we might suppose that this is the case on the Continent. In fact, from a European standpoint, these three species can be described as 'Atlantic' and are restricted to western Europe. Although Watson's classification does not cover Ireland, the same weakness remains. For example, willow-leaved inula (*Inula salicina*) occurs on the limestone shores of Lough Derg, in northern Tipperary and south-east Galway, and nowhere else in the British Isles, but on the Continent it is widespread from Greece to Norway, and from western France to the Soviet Union.

At about the same time as Watson was devising his classification scheme, Edward Forbes produced a classification of range into five elements or sub-floras, which he thought had resulted from five separate invasions into the British Isles, under different climatic conditions, and at different times. For example, Forbes thought that the Germanic flora spread into the British Isles in the 'post-glacial' period, the Scandinavian flora during the 'glacial period', and the Iberian flora (what we would now call Lusitanian) in the 'pre-glacial'. Each invasion was across a land-bridge, some of which certainly did exist, while others were figments of his imagination. The southern part of the North Sea was the land-bridge for his Germanic invasion, and dry land (much of it swampy) certainly did exist in this region during periods of low sea level. The land-bridge between Iberia and Ireland, extending as far out into the Atlantic as the present-day Sargasso Sea, does not stand up to much geological inspection. The important point about Forbes' classification is that it provoked biogeographers into thinking about the relationship between the British and European floras, which was an important step forward. Forbes' five sub-floras were as follows:

1 *Iberian or Asturian* – species found in the north of Spain;
2 *Armorican or Gallican* – species of the Channel Isles and west France;
3 *Kentish* – species of the north and north-east of France;
4 *Scandinavian or Boreal* – species of the arctic and sub-arctic;
5 *Germanic* – species of central or west-central Europe.

In the early 1930s, range classifications were published for plants (Salisbury 1932) and beetles (Deville 1930). However, during this period by far and away the most comprehensive classification of range was developed by Matthews (1937), who divided the 1,500 or so species of the British flora (excluding aliens, endemics, and microspecies) into fifteen geographical elements. Matthews, like Forbes before him, thought that since the flora of the British Isles is essentially immigrant, and apparently derived from various parts of Europe, it should be possible to analyse its components (*elements*) and indicate possible paths of invasion – an idea not unlike that of Forbes nearly a century before.

Table 7.1 Floristic elements in the British Isles according to Matthews.

Element	No. of species
1. Wide – species found at least throughout the northern temperate region.	205
2. Eurasian – species found generally distributed throughout Europe and temperate Asia	480
3. European – species generally distributed throughout Europe	130
4. Southern – species whose continental range is predominantly more southerly than the British Isles	
(a) Continental southern element – species of south and central Europe	127
(b) Oceanic west European element – species of the Atlantic fringe of Europe	76
(c) Oceanic southern element – species of south and west Europe	74
(d) Mediterranean element – species whose range is centred on the Mediterranean	38
5. Northern – species whose range is predominantly more northern than the British Isles	
(a) Continental northern element – species with a central and north European range	91
(b) Oceanic northern element – species of north-west Europe but some in north-east America	26
(c)Northern Montane – species of northern Europe and mountains further south	25
6. Continental element – species whose range is centred in central Europe and which stretches eastward into Asia	82
7. Arctic-alpine element – species characteristic of the arctic or sub-arctic or are exclusively alpine	
(a) Arctic-subarctic element – exclusively northern species	30
(b) Arctic-alpine element – northern species but also in southern mountains	106
(c) Alpine element – species of the central European mountains	9

Source: Good (1974) based on Matthews (1937)

Matthews' scheme is quite complex, and a summary of his classification is shown in Table 7.1. Some examples of the wider European range of selected species are illustrated in Figures 7.1–7.4. A novel use of Matthews' scheme is that of Jermy and Crabbe (1978), who used data from the *Atlas of the British Flora* and floral lists provided by Matthews (1955), to construct histograms for each geographical element at eleven sites along a north–south transect from Pembroke to Shetland (see Figure 7.5). Although the geographical ranges proposed by Matthews and others overlap, Jermy and Crabbe's histograms show quite well how the proportions in each floristic element change in an expected way. They also conducted a more sophisticated analysis of the data using principal components which, basically, group the sites in terms of all the phytogeographical elements

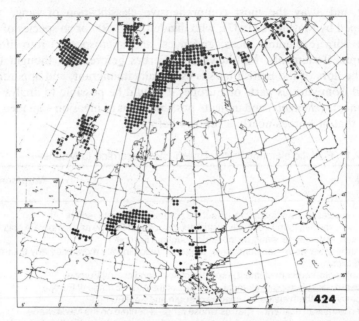

Figure 7.1 Distribution of mountain sorrel (*Oxyria digyna*) – an arctic–alpine element. (Redrawn from Jalas and Suominen [eds] 1983)

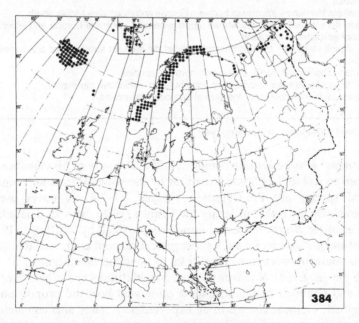

Figure 7.2 Distribution of Iceland purslane (*Koenigia islandica*) – an arctic element. (Redrawn from Jalas and Suominen [eds] 1983)

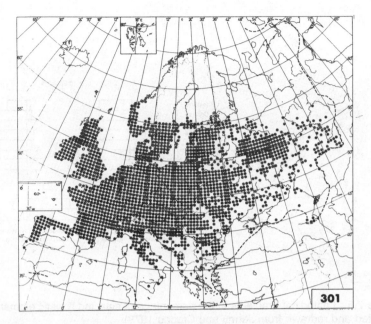

Figure 7.3 Distribution of pedunculate oak (*Quercus robur*) – a continental element. (Redrawn from Jalas and Suominen [eds] 1983)

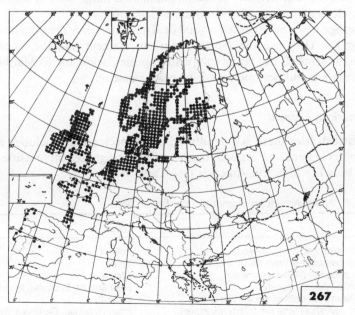

Figure 7.4 Distribution of bog myrtle (*Myrica gale*) – a northern-oceanic element. (Redrawn from Jalas and Suominen [eds] 1983)

161

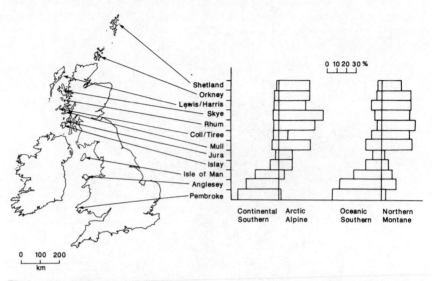

Figure 7.5 Latitudinal gradients in the percentage of selected floristic elements. (Modified and redrawn from Jermy and Crabbe 1978)

present. The results of this analysis show that Mull is near a pivotal point on the west coast of Britain between a dominance in the flora of southern-oceanic elements and northern-montane elements.

Two range types have caused a good deal of controversy over the last 100 years or so. They are the North American element and the so-called Lusitanian element, the latter comprising species belonging to Matthews' Mediterranean and oceanic/west-European elements.

The American element

In the British Isles there is a handful of plant species which belong to the American element, so named because their range centre outside the British Isles lies in North America (see Table 7.2).

Table 7.2 North American element in the British flora

Species	Location
Eriocaulon septangulare (pipewort)	Ireland
Hypericum canadense (Irish St. John's wort)	Ireland
Limosella subulata (Welsh mudwort)	Wales
Najas flexilis (slender naiad)	Lake District, Scotland, W. Ireland
Sisyrhinchium bermudiana (blue-eyed grass)	Ireland
Spiranthes romanzoffiana (Irish lady's tresses)	Inner Hebrides, Dartmoor, Ireland

Source: Webb (1983); Love and Love (1958).

There is still much speculation as to when and how this range developed. Iversen, for example, thought that the dispersal of the blue-eyed grass *Sisyrhinchium bermudiana* might be due to coastal shipping in Viking times, while Heslop-Harrison (1953) argued that recent long-distance dispersal by birds was a very probable explanation of its range. According to Godwin (1956), the key species to the enigma is the water plant, slender naiad (*Najas flexilis*) which has a long post-glacial history in Ireland and, for that matter, in western Europe. In Britain, *Najas* was already present in the Late-glacial period, and by Boreal times was widespread in the British Isles and Scandinavia. Its range could thus be more generally described as circumpolar. The fact that its present-day distribution centre is in North America, is evidently due to a severe reduction in its range in western Europe and Asia in post-glacial times. Why post-glacial changes should not have equally diminished its range in North America is not known.

Webb (1983), in discussing the American element in Ireland, suggests that some might have survived the last glacial episode on refugia off the coast of Scotland and Ireland. The fact that the pipewort (*Eriocaulon septangulare*) is present in the Gortian warm stage and the post-glacial provides strong corroborative evidence he suggests. While this may be possible, a simpler explanation is that it dispersed rapidly in the early post-glacial and its range then contracted severely.

Love and Love (1958) carried out a detailed cytological examination of the American element in the flora of the British Isles, and came to the conclusion that most species were polyploid, and that the Irish plants differ in chromosome number from most American samples. This probably indicates some speciation in a once widespread flora that has been fragmented during the Tertiary and Quaternary. Certainly, the cytological evidence falsifies Heslop-Harrison's theory. (Elkington 1984).

It is interesting to note that some common West European species have ranges which nowadays just extend into North America. These include heath grass (*Sieglingia decumbens*), lousewort (*Pedicularis sylvatica*), and the garden snail (*Cepaea hortensis*). An asymmetric range on either side of the Atlantic is not unusual and, in species which were formerly widespread, may come about simply through differences in environmental and ecological processes in the two continents. Precisely when the circumpolar flora developed is not known, but we should recall that prior to the opening of the North Atlantic Ocean there must have been considerable contact between American and European biota via land-bridges passing through Greenland.

The Lusitanian flora and fauna

The story of the Lusitanian flora and fauna can be said to have begun in 1842 when, in the woods of County Kerry, south-west Ireland, a slug was found

Table 7.3 The Lusitanian flora and fauna in the British Isles

	Ireland	England	Continent
Flora			
Arbutus unedo (strawberry tree)	W	–	Charente-Maritime (west France)
Daboecia cantabrica (St. Dabeoc's heath)	W	–	west France and north-west Spain
Erica ciliaris (Dorset heath)	–	SW	north Spain
Erica erigena (Irish heath)	W	–	west France.
Erica mackaiana (Mackay's heath)	W	–	north-west Spain (east Galicia)
Minuartia recurva (curved sandwort)	W	–	north-east Portugal
Neotinea maculata (dense-flowered orchid)	W	–	north Spain
Pinguicula grandiflora (large-flowered butterwort)	SW	–	north Spain (Asturias)
Saxifraga hirsuta (kidney saxifrage)	SW	–	north Spain (Asturias)
Saxifraga spathularis (St Patrick's cabbage)	SW	–	north-west Spain
Fauna			
Geomalacus maculosus (Kerry slug)	SW	–	Spain and Portugal
Lumbricus friendi (an earthworm)	S	–	Spain
Oritoniscus flavus (a woodlouse)	SE	–	central Pyrenees
Theba pisana (a snail)	E	SW	south-west France

Source: Webb (1983); Praeger (1939)

with a very memorable appearance; it was black with yellow spots. The discoverer had accidentally come across *Geomalacus maculosus* colloquially called the Kerry slug which is found nowhere else but Portugal. A few years later, in 1846, Forbes described similar disjunct distributions in species belonging to his Iberian sub-flora which comprised those species characteristic of south-west Ireland and whose European distribution was confined more or less to northern Spain, Portugal, and south-west France. Probably the best-known plant in this group is the strawberry tree (*Arbutus unedo*). In the last 100 years or so, a dozen or so plants and invertebrate animals have been shown to have similar disjunct distributions, and the group as a whole are called after the Roman province covering the west of the Iberian peninsula (see Table 7.3).

Since Forbes' day, biogeographers have puzzled over the Lusitanian flora and fauna. Why, for example, with a few notable exceptions do they not exist in England or north-west France? And when did the Lusitanian taxa enter the British Isles? Forbes, of course, thought that they represented a fragment of the British flora and fauna which had dispersed into Ireland and south-west England in Miocene times, across a broad land-bridge, which linked Ireland with the Iberian peninsula. Subsequently, most species had been wiped out by the 'ice-age' – or 'epoch of the northern drift' as he put it.

By the turn of the century there was an impasse. On the one hand the idea of land-based glaciers had been accepted, and Ireland was known to have been intensely glaciated. This fact alone led some, such as the famous Irish geologist Charlesworth (1930), to state quite categorically that all life in Ireland was exterminated during the glacial maxima. On the other hand there was apparently no post-glacial land-bridge connection between Iberia and Ireland over which the fauna might have dispersed, and Clement Reid's suggestion of rapid post-glacial invasion was vehemently dismissed (Praeger 1932).

Ecologically, the Lusitanian flora and fauna are quite diverse and the question arises as to whether or not any might have survived the last glaciation in Ireland. Because the Lusitanian flora and fauna have southern stations in Europe, this does not mean that they must all be considered tender and warmth-loving. In the Cordillera Cantabrica, *Daboecia* grows high enough to be covered by snow for several months of the year.

Webb (1983) suggests that one or two species, such as *Pinguicula grandiflora*, are possibly relict and survived the Midlandian glaciation, perhaps on some off-shore refuge. The present-day distribution of *Pinguicula* is in the Cantabrian mountains, the Pyrenees, the Jura, and the western Alps. Webb can envisage no staging posts for its advance northward in post-glacial times. Some species, such as *Arbutus unedo* and *Neotinea maculata* (an orchid with very tiny seeds), could have dispersed naturally, and it is worth recalling that there is a major bird-migration route between North Africa, through Spain, Scotland, and on into the Arctic. This might well explain the presence of *Neotinea* in the Isle of Man but does not explain the absence of many Lusitanian species from the British mainland.

Mitchell (1986) raises the possibility of dispersal along the west European coastal strip for species such *Aepophilus bonnairei*, a shore-dwelling bug, which can neither swim nor fly. This species has its headquarters in Morocco and Portugal but is also found on the Irish Sea coasts north to the Isle of Man, and also on the Atlantic coast of Ireland. When warmth returned at the beginning of the post-glacial, Mitchell describes *Aepophilus* as:

> marching up the French coast, skirting the embayments that then marked the English Channel and the south of the Irish Sea and making its way to the Atlantic coast of Ireland.

This might well have been the case, although dispersal by wind cannot be ruled out. If dust can be brought from the Sahara, butterflies blown in to southern England from Spain, then, surely, small bugs can also be picked up by strong southerly winds, and be deposited on the British coast.

Corbet (1962) reviews the various hypotheses for the Lusitanian fauna, and comes to the conclusion that the most plausible explanation is that they have been introduced accidentally in post-glacial times as a result of trade and sea-faring between Iberia and Ireland, perhaps dating back to 2,500 BC. As it so happens, all the Lusitanian fauna are small invertebrates, which could have easily been transported in the soil, on the roots of plants, or in animal fodder. He suggests that the flora could have entered Ireland in this way also, though he does not rule out dispersal by birds.

Before leaving our discussion of the Lusitanian flora and fauna we ought briefly to note that some species have certainly not been introduced by man, since there is fossil evidence to show that they were present in the Gortian warm stage. For example, there are fossil pollen of *Daboecia cantabrica, Erica mackaiana,* and *Erica ciliaris*, all three of which were widespread in south and west Ireland at this time. Strangely enough, the Dorset heath (*Erica ciliaris*) does not occur in Ireland today, but on the heaths of south-west England.

Endemic species

Species whose range is limited to some pre-defined area are known as endemic. Since all species must occur somewhere, the term is usually used for those species having an abnormally restricted range. Species occurring on a single island, mountain range, or perhaps country, might be termed narrowly endemic, as compared with broadly endemic species, whose range is limited to a major biogeographical region such as a continent.

Endemic species are of great interest and are often important within the context of conservation, both at the national and international level because many are rare and endangered.

Interest in endemism seems to have begun with the publication, in 1880, of Wallace's book *Island Life* in which there is a chapter on the British Isles. Wallace enlisted the help of Watson regarding possible endemics amongst the British flora, but received a rather negative response. Watson stated pretty confidently that there were no 'good species' peculiar to the British Isles. By this he meant that often there was taxonomic confusion and that in fact many species had been described elsewhere under a different name. In addition, some so-called endemic species were not worthy of that rank and had been reduced to a sub-species or, perhaps, variety by succeeding botanists. Not suprisingly, Wallace was a little disappointed with Watson's response, and in the second edition of his book, Wallace lists some 72 endemic taxa (species, sub-species, and varieties) based on information provided by Ridley.

The view of modern-day taxonomists is probably not very different from that of Watson. Not only are the taxonomic problems formidable but there is the further problem of deciding whether to include the so-called 'microspecies' of genera such as *Hieracium* and *Taraxacum*, both of which belong to the Compositae family, in which there is a good deal of non-sexual reproduction (apomixis). All apomictic reproduction gives rise to genetically identical plants called clones. For example, in the bluebell some clone patches have white flowers and others pink. Every clone is likely to be genetically different and there is the temptation to give each a separate species name.

In his assessment of British flowering plant endemics, Walters (1978) lists twenty-three species accepted in Flora Europaea, excluding all the apomictic microspecies of the genera *Rubus, Sorbus, Hieracium* and *Taraxacum* (see Table 7.4). From a statistical point of view Walters' list amounts to an endemism of about 1 per cent, as compared with 3 per cent for the Balearic Isles, and 90 per cent for the very isolated Hawaiian Isles. An assessment of endemism in the liverworts and mosses was made by Dickson (1973), who evaluated the claims of about twenty taxa regarded as endemic. In his judgment, only six species are unlikely to be discovered outside the British Isles (see Table 7.5).

The low percentage of endemism in the British vascular flora is entirely in keeping with the fact that the flora is young, and has only been isolated from Europe for a short while. The British flora is essentially re-immigrant and according to Walters, contains no clear case of palaeoendemic species which might be relict from a pre-glacial flora. In fact none of the twenty-three species listed in Table 7.4 are known to have occurred in the British Isles prior to the last glaciation. Virtually all British endemics have evolved as polyploid species, from widely distributed diploids. This has come about

Table 7.4 Flowering plant species endemic to the British Isles

Britain only	Ireland only	Britain and Ireland
Fumaria occidentalis	Salix hibernica	Euphrasia pseudokerneri (chalk-hill eyebright)
Gentianella anglica (dwarf English felwort)	Saxifraga hartii	Limonium paradoxum
Calamagrostis scotica (Scottish small-reed)		Limonium transwallianum
Alchemilla minima		Bromus pseudosecalinus
Bromus interruptus (interrupted brome)		Fumaria purpurea
Epipactis dunensis (dune helleborine)		
Euphrasia campbelliae		
Euphrasia marshallii		
Euphrasia rivularis (Snowdon eyebright)		
Euphrasia rotundifolia		
Euphrasia vigursii		
Limonium recurvum		
Primula scotica (Scottish bird's eye primrose)		
Rhynchosinapis monensis (Isle of Man cabbage)		
Rhynchosinapis wrightii (Lundy cabbage)		
Senecio cambrensis (Welsh ragwort)		

Source: Walters (1978)
note: includes only those species accepted for Flora Europaeae

Table 7.5 Possible endemic liverworts and mosses in the British Isles

Barbula reflexa var. *robusta*	*Gymnostomum recurvirostrum* var. *insigne*
Campylopus shawii	*Radula voluta*
Grimmia retracta	*Trichostomum hibernicum*

Source: Dickson (1973)

through chance mutation, which has caused some multiplication of the basic diploid set of chromosomes.

The most thorough discussion of the endemic status of the British fauna, is that of Beirne (1952), but this is now dated, and many species have undergone taxonomic revision. In the past, zoologists have been rather too ready to give species rank to taxa. This problem is particularly acute given the archipelago nature of the British Isles, where sub-species, race, and variety development is probably common on the many isolated islands. For example, according to the lists in Berry (1977) many islands in the Hebrides have their own endemic sub-species/race of field mouse (*Apodemus sylvaticus*). As he suggests, the more a species is investigated the more variety is found and the 'splitters' among the taxonomists can get to work.

The difficulties of using the older literature are highlighted in Beirne's suggestion that the red grouse is the only endemic species of bird in the British Isles; he names it *Lagopus scoticus*. Nowadays the red grouse is merely regarded as a distinct race (*Lagopus lagopus scoticus*) of the willow grouse or willow ptarmigan, whose range extends across America and Eurasia.

Genetic variation within a range

The total genetic make-up of a species is rarely uniform over its entire range and frequently a measurable geographical gradient or cline exists in some genotypic or phenotypic frequency. There may be, for example, a gradient in a species colouring (phenotypic), or perhaps a gradient in a species resistance to a pesticide (genotypic). Clines may be stable or transient, and are thought to arise in the following ways:

(i) random genetic drift;
(ii) contact between two genetically different populations;
(iii) spatially discontinuous changes of environment;
(iv) continuous environmental changes.

A cline produced by random genetic drift disappears as soon as gene flow takes place and, in terms of gradients over geographic space, we can envisage transient clines rather like ripples on the surface of a lake. Stable clines develop where the population is exposed to two opposing selection

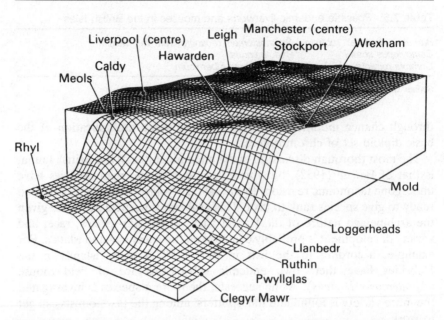

Figure 7.6 Cline of the melanic form of the peppered moth (*Biston betularia*) (From Berry 1977)

pressures at either end of its range, giving rise to a slightly different genetic make-up in the two regions. Figure 7.6 is a three dimensional plot of the cline in the frequency of the melanic (dark) form of the peppered moth (*Biston betularia*). This cline has come about due to environmental pollution. In urban areas the melanic form has a selective advantage in that it is less easy for predator birds to see it against dark polluted surfaces whereas the reverse is the case in rural areas. Berry (1977) provides a thorough account of clines and their development.

Computing floristic elements

The geographical elements described thus far have been grouped quite subjectively by eye. According to Jardine (1972) there are certain pitfalls in this approach since there is a tendency for the human eye to find groupings, even when there are none. In addition there may be an unconscious bias because of preconceptions about the factors which govern distributions. Jardine's approach to the classification of geographical elements is through numerical taxonomy using a method called cluster analysis. He collected the names of the microspecies of *Hieracium* (hawkweeds) in each cell of a 100 km grid using the distributions mapped in the *Critical Supplement to the Atlas of the British Flora* (Perring 1968). An index of similarity between each pair of species distributions was then calculated. One

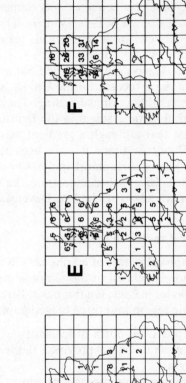

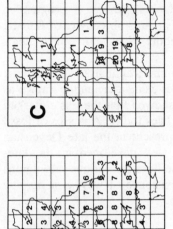
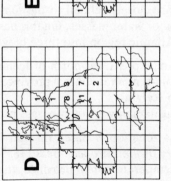

Figure 7.7 Distribution of six floristic elements (A–F) obtained by cluster analysis of *Hieracium* microspecies distributions. For each cluster the number of microspecies in each 100 km grid square is shown. (Redrawn from Jardine 1972)

appropriate measure is the number of squares two species have in common. The result of this exercise is the production of a matrix of similarity values which is then subjected to a cluster analysis. At progressively lower similarity levels this technique groups the entries in the similarity matrix. Jardine identified six groups (floristic elements) although it must be pointed out that there are many types of cluster analysis and each might produce a slightly different result. The distribution of the six floristic elements identified by Jardine are illustrated in Figure 7.7. Jardine points out that the sequence of elements A–F shows a progressive displacement along a SSE–NNW axis. The greater objectivity of floristic elements derived from numerical taxonomy does not make their interpretation any less problematical.

Birks and Deacon (1973), carried out a numerical analysis of both past and present floral records for the British Isles using a technique known as non-metric multidimensional scaling. They showed that there is a marked north–south floristic gradient in the British Isles today. Analyses of fossil data suggest that although a gradient was present in the late Devensian and mid-Flandrian times it only became well developed in the late-Flandrian. Phytogeographical analyses of the flora for various sub-regions in the British Isles also demonstrate the existence of west–east floristic gradients, particularly in the late Devensian.

Geographical ranges in the sea

The geographical ranges of marine taxa in the shelf seas around the British Isles are closely related to the temperature and salinity characteristics of the major water masses. On this basis, Briggs (1974) has divided the north-east Atlantic region into three biogeographical provinces (see Figure 7.8):

(1) Arctic province: north of Iceland;
(2) cold-temperate boreal province: the seas around the British Isles north to Iceland;
(3) Lusitanian warm-temperate province: Biscay to Gibraltar.

This zonation is based on that in Forbes' classic treatise *The Natural History of the European Seas* (1859) and differs only in the naming of the cold temperate boreal province, the southern part of which Forbes named the celtic province, which included all the British Isles except the Shetlands.

A fundamental point to make here is that the seas around the British Isles are very much a mixing zone, with strong faunal connections both to the north and the south. As both temperatures and sea levels rose in the post-glacial period, southern taxa extended their range northward – especially up the west coast of the British Isles. On the other hand many northern taxa with arctic affinities, which were associated with the seas

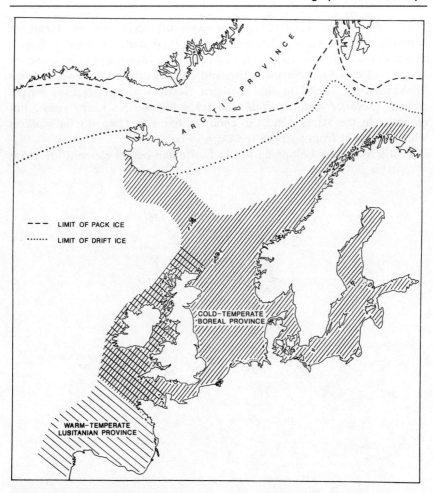

Figure 7.8 Biogeographical provinces of the north-east Atlantic. (Based on Briggs 1974 and Forbes 1859)

marginal to the British ice sheets, retreated northward and today their southern extent is limited to deeper, cold waters.

Because of the considerable complexity of the marine environment, and the position of the British Isles midway between the Arctic and warm-temperate provinces the marine geographical elements recognized by biogeographers are best thought of as mixtures. In Ekman's classification of the fauna, three elements are recognized, and we may also use his divisions for the marine flora in the British seas (Ekman 1953):

(i) Arctic-boreal element;
(ii) Lusitanian-boreal element;
(iii) Boreal element.

The Arctic-boreal component is comparatively small in terms of common species. Among those which have their southern limits in British seas are: the red seaweed (*Odonthalia dentata*), found along the shores of Northern Ireland and Northern England; the sea-urchin (*Stongylocentrotus drobachiensis*), found in the northern North Sea; the starfish, purple sunstar (*Solaster endica*), found as far south as the Kerry coast, but common in the Arctic; and the common barnacle (*Balanus balanoides*) which is absent from parts of the Cornish coast.

Southern species belonging to the Lusitanian-boreal element are more

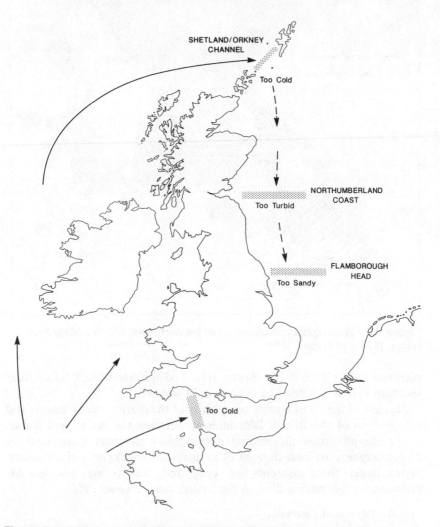

Figure 7.9 Major marine biogeographical barriers around the British Isles. (Modified and redrawn from Earll and Erwin 1983)

numerous. In the cold waters of the North Sea, their range is confined and many fail even to extend beyond the eastern English Channel. On the west coast, the Lusitanian current and North Atlantic drift extend the range of this element as far north as the Shetlands and southern Norway. Some fairly common examples in this group are: the cross-cut carpet shell (*Venerupis decussata*), which is common on the south coast; the flat oyster (*Ostrea edulis*); the barnacle (*Balanus perforatus*); and, according to Ekman, all but two of our crab species.

The boreal element is regarded separately from the mixtures of southern and northern taxa, due to the occurrence of an endemic element. Among those quoted by Ekman are: the brittle-star (*Ophiura affinis*); the common sea-urchin (*Echinus esculentus*); and a considerable number of fish species including the dab (*Pleuronectes limanda*) and the greater sand-eel (*Ammodytes lanceolatus*). Levels of endemism in the boreal element are quite high. Briggs (1974) suggests that about 23 per cent of echinoderms and 28 per cent of fish species are endemic.

Earll and Farnham (1983) recognize a number of important range boundaries for sublittoral taxa (see Figure 7.9). For the Lusitanian-boreal element, the English Channel is an important boundary to the colonization of the west European coasts and the eastern English Channel. Along the west coast of Scotland, southern Lusitanian-boreal species are quite common, but there is a major range boundary at the Shetland–Orkney channel, reflecting the presence of colder water. Once into the North Sea

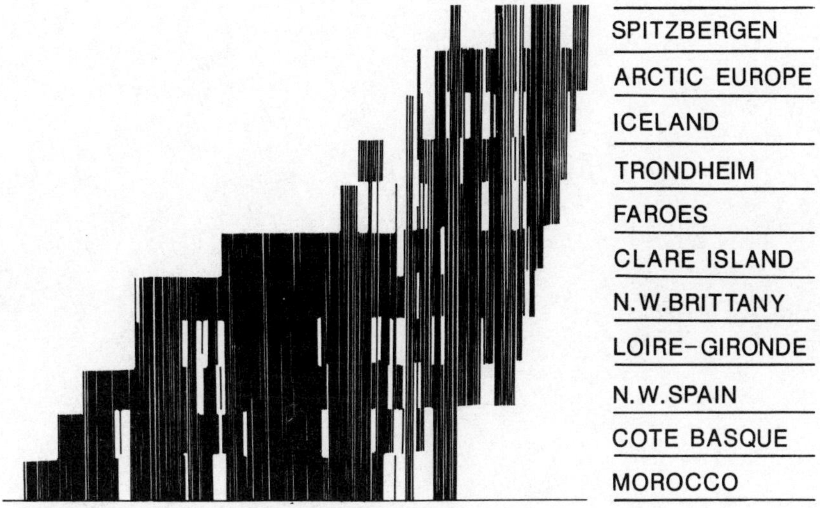

Figure 7.10 Latitudinal ranges of benthic algae on north-east Atlantic coasts. Each line represents one species. (Modified and redrawn from Hoek and Donze 1967)

by this northern route, there are coastal range boundaries along the Northumberland coast due to a marked increase in turbidity, and at Flamborough Head, beyond which there is little rocky substrate exposed. The diversity of Arctic-boreal taxa is similarly reduced along the North Sea coast, though the picture is not so clear.

The latitudinal range of benthic algae on the north-east Atlantic coasts has been described by Hoek and Donze (1967). They examined the species occurrence at eleven coastal sites, from Morocco northward to Spitzbergen. Although there is a gradual replacement of southern species by northern species the flora only changes gradually between Morocco and north-west Brittany. Between Brittany and the Faroes, however, there is a sharp decrease in the number of species present. This is shown diagrammatically in Figure 7.10. In this figure, each vertical line represents the latitudinal range of one species. The floristic discontinuity between the flora of Clare Island – off the west Irish coast – and that of the Faroes is very clear, and emphasizes again the way in which the British seas are an important mixing zone between taxa whose ranges are essentially southern, and those which are essentially northern.

Chapter eight

Environmental change

In order to understand how biogeographical ranges become established we need to consider the major environmental changes that have occurred over geological time. It is of course possible to consider the whole of the geological column, but for the purposes of this chapter a natural starting point is the Tertiary Period, whose fauna and flora have recognizable, modern-day, equivalents.

The Tertiary back-cloth

At both the generic and specific level, there are affinities between the North American and north-west European fauna and flora, and to understand how this came about we need to delve a little into the world of plate tectonics and the evolution of the North Atlantic Ocean. In another sense, the Tertiary is a good starting point, since it also marks the end of the Mesozoic Era and the beginning of the Cenozoic Era (see Table 8.1). At the end of the Cretaceous Period many groups of organisms became extinct. On land, the palm-like *Cycads* had more or less died out, the great reptiles that had dominated the world in Mesozoic times vanished abruptly, and mammals began to establish themselves widely (Pomerol 1982).

At the start of the Tertiary, southern Britain lay about 40° N, and together with the rest of Europe was steadily drifting northward. By the late Miocene, the area lay at nearly 50° N. This northerly drift into higher latitudes helped produce pronounced changes in climate, from tropical in the early Tertiary, to the temperate regimes of the late Tertiary and the present day.

The Palaeogene

The opening of the North Atlantic Ocean dominates the early Tertiary biogeography of the British Isles. In the preceding Cretaceous Period, shallow epicontinental seas bisected North America and Asia, resulting in

Biogeography of the British Isles

Table 8.1 Principal divisions of the Cenozoic Era

Era	Period		Epoch	Radiometric age (Ma)	
C	QUATERNARY		Holocene	0.01	
E			Pleistocene		
N			----------	1.8	
O		N	Pliocene		
		e	----------	5.0	
Z		o	Miocene		frosts begin
		g			
O		e			
		n			
I	TERTIARY	e	----------	22.5	
		P	Oligocene		warm temperate – less humid
C		a			
		l	----------	37.5	
		a	Eocene		
		e			
		o	----------	55.0	
		g			tropical – humid
		e	Palaeocene		
		n			
		e	----------	65.0	

two land-masses, one comprising Europe and eastern North America, the other Asia and western North America. At this time, the flora and fauna of Europe and east North America were very similar and there were no salt-water barriers to deter dispersal. The present-day taxonomic affinities on either side of the Atlantic Ocean are part of a generalized track, in the sense of Croizat, the construction of which provides cartographic evidence for vicariance induced by plate tectonic movement (Mckenna 1975).

By the late Cretaceous and early Tertiary Period the sea-floor spreading along the Mid-Atlantic Ridge began to develop an extension of the Atlantic Ocean between Greenland, on the one side, and Norway and the British Isles, on the other. During its early development it was not a totally effective salt-water barrier to floral and faunal exchange, because of the existence of two land-bridges joining Europe and North America, both of which passed through Greenland (see Figure 8.1) Both land-bridges were probably in existence throughout the Palaeocene and early Eocene. To the south, joining the British Isles (and hence Europe), the Faroes, Iceland and Greenland, lay the Thule land-bridge (Thule of classical times was probably the Shetlands), which is thought to have been for the most part sub-aerial. To the north lay the De Geer land-bridge across the Barents Shelf, through Svalbard, northern Greenland and on into Ellesmere Island.

178

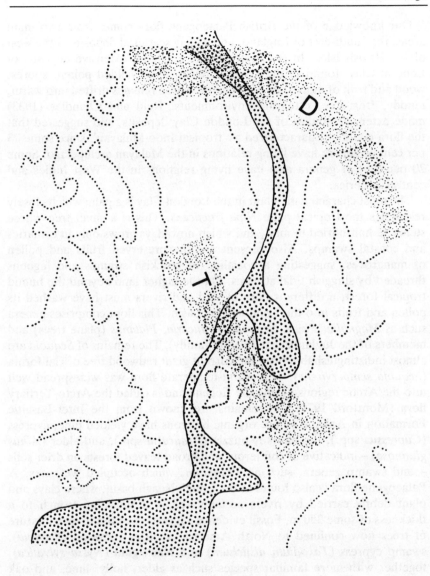

D – De Geer Land-Bridge

T – Thule Land-Bridge

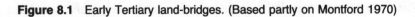 – Marine Troughs and Basins

Figure 8.1 Early Tertiary land-bridges. (Based partly on Montford 1970)

Our knowledge of the British Palaeogene flora comes from two main areas: the south-east of England, and widely scattered deposits in the west of the British Isles. In the south-east of England, extensive deposits of London Clay, found in the London Basin, have yielded pollen, spores, wood and fruit of more than 350 species, whose remains drifted into warm, muddy, frost-free, estuarine environments. Reid and Chandler (1933) made extensive studies of the London Clay deposits, and suggested that the flora could be characterized as 'tropical Indo-Malayan', since some 73 per cent of genera have living relations in the Malayan archipelago. Some 20 per cent of genera also have living relations in the West Indies and central America.

The most characteristic plant in the London Clay is a palm which closely resembles the tropical palm *Nipa fructicans*. This is a small tree whose stems lie half-buried in mud, and which nowadays grows only in estuaries and coastal swamps. Also present are the preserved fruits and pollen of mangroves, suggesting a coastline of brackish swamps and lagoons threaded by sluggish tidal streams. On the higher land beyond the humid tropical forest, a different flora existed and rivers must have washed its pollen and fruits into the London Clay basin. This flora comprises genera such as *Magnolia, Pinus, Sequoia, Araucaria, Platanus* (plane trees) and members of the Juglandaceae (walnut family). The remains of *Sequoia* are almost indistinguishable from the current giant redwood tree of California (*Sequoia sempervirens*). This warm-temperate flora was widespread well into the Arctic regions during the Eocene and is called the Arcto-Tertiary flora (Montford 1970). Its remains are known from the Inter-Basaltic Formation in Antrim, whose organic horizons have yielded pine, cypress (*Cupressus* spp.), the monkey-puzzle (*Araucaria* spp.), and alder (*Alnus glutinosa*) – indicative of coniferous and broad-leaved forests on drier soils – and swamp genera, such as *Taxodium*, which occupied wet sites. A Palaeogene flora is also found in the Lough Neagh basin, where clays and plant debris carried by rivers accumulated in a subsiding trough to a thickness of some 350 m. Fossil evidence suggests the flora to be a mixture of trees now confined to North America, such as redwood (*Sequoia*), swamp cypress (*Taxodium distichum*) and black gum (*Nyssa sylvatica*), together with more familiar species such as alder, holly, lime, and oak (Boulter 1980).

Throughout the rest of the Tertiary, the story is one of a progressive extinction of Palaeogene species, as the climate became cooler and more variable. In the late Eocene the temperature of the North Sea fell dramatically as a result of a major climatic deterioration, and thermophilous species declined in importance. The palms, for example, require a frost-free environment and became extinct during the Miocene when frost became a climatic feature of the lowlands.

The early Tertiary terrestrial fauna is not so well known as the flora but there must have been a considerable amount of traffic across the Thule land-bridge into the British Isles, and on into Europe. Snakes, crocodiles, and mammals of South American ancestry occur in the west European Eocene, as does the only known anteater to have lived in Europe (*Eurotamandua joresi*) which is found in middle Eocene lake deposits in north-west Europe. It dispersed from South America in the late Cretaceous, and entered Europe in the early Tertiary. The Thule land-bridge evidently foundered soon after the Eocene, and, since that time, the West European biota has become progressively dissimilar to those of the surrounding continents (Hoch 1982).

The Neogene

At the beginning of the Neogene, the British Isles were subject to major orogenic uplift, as Europe reeled under the intense compressional forces brought about by the northward drift of the African plate. The major topographical features of the present-day British landscape essentially date from this period. The climatic effects of this uplift must have been additional to global changes and, in the newly formed uplands, frost and cool, cloudy winters were common features of the environment. Terrigenous Miocene sediments did not accumulate in Britain because of extensive erosion, and our knowledge of the Neogene flora and fauna is based only on Pliocene deposits. In north-west Europe the Pliocene floral remains indicate an oceanic type of climate. A system of depressions, tracking across a large, Atlantic water-body must have been fully established at this time.

In Ireland, only one Neogene deposit has so far been discovered – at Hollymount, north of Carlow, in south-east Ireland (Watts 1985). Pollen referable to pine (*Pinus* spp.), Japanese umbrella pine (*Sciadopitys verticillata*), sweet gum (*Liquidambar styraciflua*), birch (*Betula* spp.), holly (*Ilex aquifolium*), hazel (*Corylus avellana*), and hornbeam (*Carpinus betulus*) has been identified. Hemlock (*Tsuga* spp.) is also present for the first time in the flora, and the remains of heather (*Erica* spp.) and rhododendron (*Rhododendron* spp.) have been found. Such a mixed floral assemblage suggests a climate slightly warmer and less seasonal than at present (Mitchell 1976).

In England, fossiliferous Pliocene deposits are known from East Anglia and from central Derbyshire. In East Anglia, pollen analyses of marine Coralline Crag clays at Orford, on the Suffolk coast, show the flora to have contained a variety of mixed woodland genera comprising conifers such as *Pinus* (pines), *Picea* (spruces), *Abies* (firs), *Tsuga* (hemlocks), and broadleaves such as *Betula* (birches), *Ulmus* (elms), *Quercus* (oaks), *Liquidambar* (sweet gums), *Carpinus* (hornbeams), *Pterocarya* (wing nut), *Acer* (maples), and *Ilex* (hollies). West (1980b) suggests this flora represents an 'upland' element, in contrast to the presence of *Sciadopitys, Sequoia, Taxodium, Cupressus*, and *Alnus*, which were associated with poorly-drained, 'lowland' environments. A similar flora has been recorded at Brassington, in Derbyshire, where Neogene sediments accumulated in karstic depressions (Boulter 1971).

A climatic interpretation of the Pliocene floras of Europe is given by the Polish botanist, Szafer, who zoned Europe into three regions: an outer zone north of the Alps, Carpathians and the Pyrenees; a south-eastern zone in the Balkans; and a southern zone around the Mediterranean (Harris 1950). For the outer zone, the Pliocene floras indicate a January mean temperature of about 6 °C, a July mean of about 25 °C, and a mean annual temperature of 16 °C. Annual rainfall would have been twice that of today. In the British Isles, on the western edge of the outer zone, temperatures were probably a few degrees lower.

Toward the end of the Pliocene, there was a major climatic deterioration heralding the beginning of the Quaternary Period, with its attendant glacial episodes. In East Anglia, *Sciadopitys, Sequoia, Taxodium*, and *Liquidambar* became extinct. In a broader context, such floristic changes were part and parcel of the great modification and fragmentation of the once-widespread late-Tertiary circumboreal, mixed mesophytic forests.

Upper Pliocene faunal remains are preserved in the Red Crag Nodule bed which crops out in east Suffolk. Among the taxa recorded are: porcupine (*Hystrix*); beaver (*Castor fiber*); a bear-like carnivore (*Agriotherium*); an extinct hyena (*Hyaena peerrieri*); an extinct panda (*Parailurus anglicus*); an elephant-like mastodont (*Zygolophodon borsoni*); true one-toed horses (*Equus* spp.); an extinct three-toed horse (*Hipparion*); and an extinct rhinoceros (*Dicerorhinus megarhinus*). Marine vertebrate remains include whales, walrus, fishes, and turtles (Stuart 1982).

In Pliocene times the eastern end of the English Channel was mostly dry land, and European fauna were able to disperse more or less unhindered into southern England. Pomerol (1982) describes the north European fauna as follows:

in the prairies of grasses and legumes were herds of mastodons, deinotheriums and anchitheriums (all three are types of primitive elephant-like mammal with a trunk and tusks). The oak forests were frequented by pigs and wild boar. Tapir and rhinoceros wallowed in the marshlands. Squirrels and monkeys climbed in the trees. Beavers built dams in the still waters. A rich fauna of insects supplied the table of insect-eaters. For the hyenas, civets and cats (*Machairodus* – a large tiger-like cat with large canine teeth) the woods and prairies provided an abundance of prey.

At the end of the Pliocene and the beginning of the Quaternary many large mammals became extinct, though genera such as *Equus* and *Elephas* which had only just appeared, are alive to the present day.

The Quaternary Period

General pattern of events

The Quaternary Period is characterized by an alternation of cold and temperate climatic episodes. During cold episodes, flora and fauna migrated to refugia, mainly in mainland Europe, to await the return of warmer conditions, when some were able to disperse back into the British Isles. This repeated, forced movement of the British flora and fauna over obstacles such as water bodies and mountain ranges has led to its progressive impoverishment, as compared with that of Europe. During the Lower and Middle Pleistocene it is thought that at least the eastern end of the Straits of Dover remained land, across which the fauna and flora could migrate. It was probably only during the Upper Pleistocene temperate stages, that the Channel became a formidable barrier to dispersal.

The climatic amelioration at the end of a cold episode, its progression to an optimum, and its subsequent retroregression at the onset of the next cold episode, was mirrored by a vegetation cycle with accompanying changes in the fauna and soils. Evidence for the changes in the flora is provided by their pollen grains and spores, which are well preserved in water-logged environments, such as peat bogs. The microscopic appearance of pollen grains reveals details of pores, furrows, and surface sculpturing which allow the palynologist to identify them occasionally to species, but more generally to genus or family level (see Figure 8.2). Characteristic assemblages of pollen in sediments (pollen assemblage zones), may be highly correlated with, for example, climate or human activity. A pollen zone is a biostratigraphic unit (a *biozone*) and can be defined as a body of sediment containing a characteristic and homogeneous content of pollen and spores.

In the British Isles, one widely used biozonation has been that of Godwin (1940) who described eight pollen zones and sub-zones for the Holocene (Flandrian) Stage of England and Wales. These, with the addition of three zones for the Late Devensian Stage are shown in Figure 8.3 together with the descriptive units (Boreal, Atlantic, etc.) of the Scandinavian botanists, Blytt and Sernander. Strictly speaking, the Godwin zonation is difficult to apply fully both in Ireland and Scotland but will do for our purposes here. Nowadays, many palynologists prefer to work with local and regional pollen assemblage zones rather than the broad Godwin zonation. West (1970) provides a discussion of this

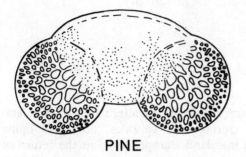
PINE

ASH

ALDER

BIRCH

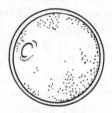
GRASS

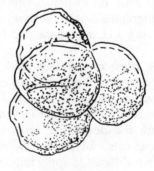
HEATHER

Figure 8.2 Microscopic appearance of some common pollen grains

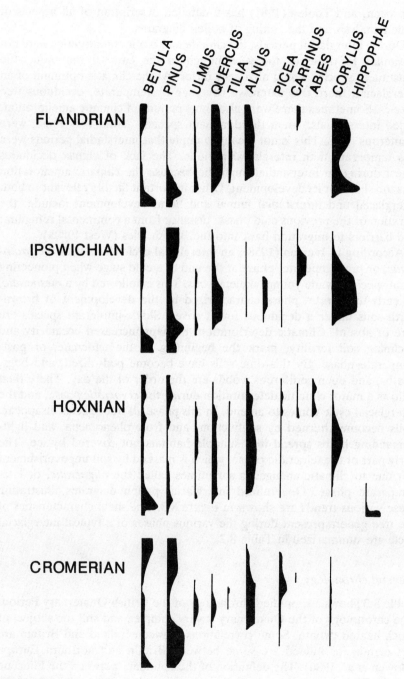

Figure 8.3 Generalized interglacial pollen diagrams. (Modified and redrawn from Bowen 1978)

approach, and Tooley (1981) has a detailed description of all aspects of pollen analysis and the zoning of pollen diagrams.

During interglacial periods, the climate and biotic communities were not dissimilar to those of today, although in the case of the biota, the constituent species would obviously differ. At the climatic optimum of an interglacial, there was extensive coverage by temperate, deciduous tree species. Sometimes there were also short periods of climatic amelioration, called interstadials, when the dominant species, at the very most, were coniferous trees. This is not meant to imply that interstadial periods were less temperate than interglacial periods. The lack of climax deciduous forest during an interstadial might be because the climatic amelioration was too short for its development. Other important factors relevant to both interglacial and interstadial faunal and floral development include: the duration of the previous cold phase; distance from a continental refugium; and barriers to migration back into the British Isles (West 1980a).

According to Iversen (1958), an interglacial cycle starts with the *proto-cratic*, or pre-temperate, phase at the end of a cold stage when pioneering plant species invade young, skeletal soils. This is followed by a *mesocratic*, or early-temperate, phase characterized by the development of brown-earth soils under a deciduous forest cover. Shade-intolerant species are rare or absent. Climatic deterioration, perhaps increased oceanicity and declining soil fertility, mark the beginning of the *telocratic*, or post-temperate phase. By this time soils have become podsolized, and bogs, heaths, and open coniferous woods are the order of the day. There then follows a major climatic deterioration during the *cryocratic* phase, and the interglacial cycle comes to an end. In this phase all tree species disappear. Soils become churned by solifluction, and frost phenomena, and light-demanding herbs spread into suitable habitats not covered by ice. The early part of the telocratic phase, which is marked by soil impoverishment not due to climatic changes is sometimes called the *oligocratic*, or late-temperate phase. Generalized interglacial pollen diagrams illustrating these various trends are shown in Figure 8.3. The main characteristics of the tree genera present during the various phases of a typical interglacial cycle are summarized in Table 8.2.

General chronology

Table 8.3 shows a simplified chronology of the British Quaternary Period. The chronology of the Quaternary is very complex and still the subject of much heated debate. Some correlations between Ireland and Britain are not certain, as indeed are some between Britain and northern Europe (Bowen *et al.* 1986). The definition of the boundary between the Pliocene and the Pleistocene depends on the criteria used. In the lower part of the Red Crag Formation, above the Pliocene basal conglomerate (Red Crag

Table 8.2 General characteristics of important tree genera during an interglacial cycle

Characteristics	Protocratic	Mesocratic	Oligocratic
	Betula	Quercus	Picea
	Populus	Ulmus	Abies
	Salix	Tilia	Fagus
Migration rate (m / year)	>1,000	500–1,000	<500
Competitive tolerance	low	high	high
Seedling tolerance	light-demanding	shade-tolerant	shade-tolerant
Growth-rate	fast	slow	slow
Soil preference	fertile	brown earth	podsol
Life-history traits	r-selected	K-selected	?
Ecological traits	ruderal	competitive	stress-tolerant

Source: Birks (1986)

Nodule Bed), cold water and arctic molluscs appear for the first time in the stratigraphic record. Species such as *Neptunea despecta*, *Buccinum groenlandicum*, and *Serripes groenlandicus*, indicate an invasion of the North Sea by northern geographical elements and the onset of cold conditions at the beginning of the Quaternary. This cold episode is equated with the Pre-Ludhamian stage, also known as the Waltonian stage by some authors.

Lower Pleistocene floras and faunas

Practically all the details of the Lower Pleistocene floras and faunas have been obtained from the marine-crag sediments on the East Anglian coast, the exception being the faunal remains in the Dove Holes cave in Derbyshire. The stratigraphic sequence shows an alternation of temperate and cold stages and, although there is no direct evidence that the early cold stages were glacial in nature, there is substantial indirect evidence in the form of far-travelled erratics in the marine and river deposits of the Thames valley and East Anglia as early as the Baventian stage (Bowen *et al.* 1986). In Table 8.3 the relationship between the Bramertonian and Baventian should not be thought of as known with certainty since there are no sites where datable deposits exist one upon the other. Bowen *et al.* (1986) suggested that the Baventian is, in fact, the younger of the two stages which implies a number of gaps in the known stratigraphy. Little is known about the biota of these early cold stages, each of which lasted considerably longer than the intervening warm stages.

Temperate stages

Lower Pleistocene temperate stages are characterized by species forming a mixed coniferous–deciduous forest but few tree genera present in the

Table 8.3 Stages in the Quaternary of the British Isles

Epoch		Stage		
		Britain (cold temperate)		Ireland (cold temperate)
Holocene		Flandrian		Littletonian
		-----------------------------------	10,000 BP	-------------------
P L E I S T O C E N E	Upper	Devensian (G) Ipswichian Wolstonian (?) (G)	110,000	Midlandian (G) ? Munsterian (G)
	Middle	Hoxnian Anglian (G) Cromerian Beestonian (G)	350,000	Gortian Pre-Gortian (G) ? ?
	Lower	Pastonian Pre-Pastonian (G) Bramertonian Baventian (G) Antian Thurnian Ludhamian Pre-Ludhamian	2,000,000	? ? ? ? ? ? ?

Sources: Stuart(1982); Mitchell (1976)
Notes: (G) indicates a cold stage and the development of ice sheets. Parts of the Lower Pleistocene sequence are disputed as is the presence of the Wolstonian stage in the Upper Pleistocene

late Tertiary flora survived the Pre-Ludhamian cold stage. *Sciadopitys, Sequoia, Taxodium*, and *Liquidambar* all became extinct in the British Isles. One genus that did survive was the hemlock spruce (*Tsuga*), now found in eastern Asia and North America, but extinct from Europe. It is a very characteristic tree of these early temperate stages. By the Pastonian very little *Tsuga* was present in the British Isles and soon after it, too, became extinct.

The faunas of the early temperate stages were characterized by several species of voles belonging to the genus *Mimomys*, comb-antlered deer (*Eucladoceros* spp.), the zebra-like Steno's horse (*Equus stenonis*), and the Auvergne mastodont (*Anancus arvernensis*). The mastodonts had straight tusks and superficially looked rather like short-limbed elephants. For the most part mastodonts had become extinct at the end of the Pliocene, and in early Pleistocene ecosystems their niche was probably filled by members of the elephant family which were evolving and expanding their range from centres in Africa and Eurasia (Nilsson 1983). One such elephantid (true elephant) was *Archidiskodon meriodonalis* – a primitive form of mammoth which, to judge by its rather poorly developed grinding teeth, might have been adapted to woodland life rather than a diet of hard-chewed grasses.

Cold stages

The floras of the Lower Pleistocene cold stages are characterized by species of heath, such as the crowberry (*Empetrum nigrum*), together with pine, birch, and alder, suggesting an open park-tundra vegetation. Many of the previously mentioned temperate-stage fauna are also present in the intervening cold stages. Stuart (1982) also notes the presence of the extinct gallic elk (*Alces gallicus*) in the Pre-Pastonian. Although smaller than its modern-day descendants in Scandinavia and North America, the gallic elk's antlers with their long beam and small palmation, still had a spread of about 3 metres and it presumably frequented open country rather than forest, thus confirming the floral record.

During both the cold and temperate stages of the Lower Pleistocene, a land isthmus connected southern England to Europe and both the Thames and the Rhine drained north. Presumably, this isthmus became broader during the cold stages, when global sea levels fell. We have no way of knowing if land connections existed between Ireland and Britain at this time.

Middle Pleistocene floras and faunas

During the Beestonian and Anglian stages, the British Isles were extensively glaciated. No glacial tills of Beestonian age have been found, but vast amounts of fluvioglacial terrace sediments accumulated in the Thames valley. During the Anglian glaciation, ice spread far south in Britain reaching north London, the Bristol Channel, and the north Cornish coast. At this time the southern North Sea basin was occupied by a large proglacial lake, whose westward drainage eroded the isthmus at the eastern end of the Channel. In Ireland, Mitchell equates the Anglian glaciation with a Pre-Gortian cold stage, and suggests that the basal sandy clays underlying Gortian fossiliferous muds, near Gort in County Galway, date from this stage. From a glaciological point of view, it might be argued that if glaciers reached far south in England, then Ireland must have also been extensively glaciated. The floras of the Hoxnian cold stage and the Holsteinian interglacial in north-west Europe are very similar, which suggests that there were few barriers to plant migration into Britain at this time. Whether or not Ireland was isolated by a water barrier is not known.

Temperate stages

The Cromerian and Hoxnian interglacial stages in Britain and the Gortian interglacial in Ireland show broadly similar cycles of vegetation development along the lines of the four phases identified by Iversen.

The flora of the protocratic and cryocratic phases of the Cromerian stage comprised herbaceous species, together with birch and pine. During the mesocratic phase there was a mixture of thermophilous deciduous broadleaf tree species, together with spruce (*Picea abies*) and common silver fir (*Abies alba*). The presence of the silver fir indicates an oceanic climate possibly warmer than at present. There are several differences in the floral composition between the protocratic phases of Cromerian and the preceding Pastonian stage. In the Cromerian there is sizable representation of *Abies* and *Tilia*, a delayed appearance of *Carpinus*, and the distinct absence or scarcity of *Hippophae rhamnoides* (sea buckthorn). With the exception of spruce, silver fir, *Trapa natans* (water chestnut), *Corema album* (a heath belonging to the family Empetraceae), *Najas minor* and *Najas tenuissima* (both aquatic plants), which are all now confined to mainland Europe, the rest of the 150 or so species of the Cromerian flora still occur in the British Isles (Nilsson 1983).

In spite of relatively high sea levels, as compared with the present temperate stage, Britain appears to have been connected to mainland Europe throughout the Cromerian and Hoxnian interglacial stages and fauna had little difficulty in entering Britain. The actual nature of the land-bridge during these stages is difficult to reconstruct because the area is one of tectonic subsidence and the palaeogeography is complex. It should also be noted that even if the land-bridge was partially inundated, the seaway would certainly not be as wide as, say, the present-day Straits of Dover which have been progressively opened by both fluvial and marine erosion throughout the Quaternary. Important mammal taxa from the Cromerian and later interglacial stages are listed in Table 8.4. The presence of large

Table 8.4 Some British interglacial mammals

	Cromerian	Hoxnian	Ipswichian	Flandrian
Alces latifrons (extinct elk)	+	–	–	–
Bos primigenius (auroch)	–	+	+	+
Crocuta crocuta (spotted hyaena)	+	–	+	+
Desmana moschata (Russian desman)	+	+	–	–
Dicerorhinus etruscus (extinct rhinoceros)	+	–	–	–
Equus spp. (a caballine horse)	+	–	–	–
Hippopotamus spp.(hippopotamuses)	+	–	+	–
Macaca spp. (macaque monkey)	+	+	–	–
Mammuthus primigenius (mammoth)	–	–	+	+
Megaceros giganteus (giant deer)	–	+	+	–
Megaceros verticornis (giant deer)	+	–	–	–
Mimomys savini (extinct primitive water vole)	+	–	–	–
Palaeoloxodon antiquus (straight-tusked elephant)	+	+	+	–
Panthera leo (lion)	+	+	+	–
Ursus arctos (brown bear)	–	–	+	+

Sources: Stuart (1982), Nilsson (1983)

deer carrying immense outspread antlers suggests that the Cromerian woodlands were not very dense.

The flora of the Hoxnian interglacial stage, which lasted about 20,000 years, is broadly similar to that of the Cromerian, although there are differences in the performance of some species. West (1980a) notes that the pollen frequencies of *Pinus* are notably less in the Hoxnian than in other Quaternary interglacial stages. This low frequency, particularly in the protocratic and early mesocratic stages is almost certainly due to pine suffering competition from deciduous trees on the calcareous soils formed by the great spread of chalky glacial tills laid down by Anglian ice-sheets as they crossed East Anglia.

In Ireland, the Gortian interglacial produced a remarkably rich flora. Many species, such as Rhododendron (*Rhododendron ponticum*), Mackay's heath (*Erica mackaiana*), the Dorset heath (*Erica ciliaris*), and St Dabeoc's heath (*Daboecia cantabrica*), did not occur in Britain at this time. Today, these heath species have their main distributions in south-west France, northern Spain, and Portugal.

The Gortian heath species are part of the Lusitanian element (see Chapter seven), whose present-day disjunct distribution is so puzzling. Evidently, this group also had a disjunct distribution in the Gortian which suggests that the disjunction developed by natural means, rather than by the intervention of man. Although palaeolithic man was in England during the Hoxnian, there is no evidence of man in Ireland during the Gortian. A number of species now confined to North America are also present in the Gortian flora. The water fern *Azolla filiculoides* and *Eriocaulon septangulare* are two examples.

Mitchell (1976) has attempted to construct a schematic pollen-diagram to show the development of woodland during the Gortian stage (see Figure 8.4). The forest flora during the mesocratic phase comprised oak, with smaller amounts of alder, hazel, and ash, together with the evergreens, holly and yew (*Taxus baccata*); elm was relatively scarce. In the telocratic phase, soil acidity encouraged the appearance of *Abies*, *Picea*, and *Rhododendron*. In some Gortian deposits, the spores of filmy ferns (Hymenophyllaceae) are recorded. These ferns are epiphytic on boulders and trees in oceanic environments and provide strong evidence for a high-rainfall, oceanic climate, when sea levels were as much as 25 m above today's level.

There are no vertebrate fossil records of Gortian age so far discovered. This is not to say that terrestrial vertebrates did not inhabit Ireland at this time. A more likely explanation is that their remains have been destroyed by extensive Upper Pleistocene glaciations. The vertebrate faunas of the Hoxnian were apparently different in several respects from their European Holsteinian counterparts. In Europe spotted hyenas and hippopotamuses were absent, while in the British Hoxnian, bison, buffalo, and mammoth

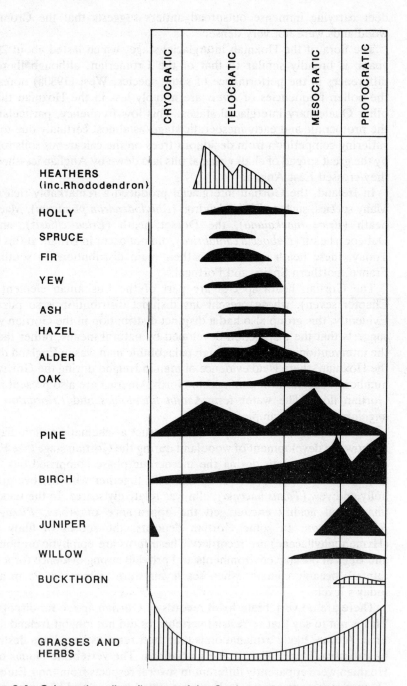

Figure 8.4 Schematic pollen diagram of the Gortian warm stage in Ireland together with Iversen's interglacial phases. (Based partly on Mitchell 1976)

fossils are lacking. These differences may represent genuine geographical variation perhaps due to climate. On the other hand, it should be said that there have not been all that many finds of Hoxnian fossils, and the picture may be very incomplete.

Cold stages

The flora of the Beestonian and Anglian cold stages are not well known but it appears that they were very similar to those of the last, Devensian, cold stage, which has been studied in some detail. West (1980b) examined the pollen of Beestonian marine and freshwater sediments on the East Anglian coast, and concluded that the floras show the presence of a herb-rich grassland, with local shrub communities containing some birch and small-leaved willow species (*Salix* spp.). A number of phytogeographical elements were present, and they suggest a steppe-tundra aspect to the vegetation. They include:

(i) Arctic-alpine/arctic-mountain elements
    dwarf birch (*Betula nana*)
    mountain sorrel (*Oxyria digyna*)
    purple saxifrage (*Saxifraga oppositifolia*)
(ii) Alpine elements
    mossy cyphel (*Cherleria sedoides*)
(iii) Maritime elements
    thrift (*Armeria maritima*)
    sea plantain (*Plantago maritima*)
(iv) Ruderal elements
    common orache (*Atriplex patula*)
    perennial flax (*Linum anglicum*)
    field mouse-ear chickweed (*Cerastium arvense*)
    mugwort (*Artemisia spp.*)

West suggests that such floras result from the influence of the cool North Atlantic, producing severe, long winters and short, sharp summers promoting herbaceous species but severely restricting tree growth. The lack of competition from trees, the great diversity of slope, aspect, and soil created a diversity of habitats. On slopes, active gelifluction would favour ruderal species. Maritime elements, with their requirement for halophytic conditions, survived where the migration of water to form ground ice had left behind salt-enriched soils. The presence of periglacial phenomena, together with warmth demanding species, suggests markedly continental climatic conditions.

Not much is known about the fauna of Beestonian and Anglian stages. Any animals present grazed on herb-rich meadows, underlain here and there by sporadic permafrost. The remains of bison, red deer (*Cervus elaphus*), reindeer (*Rangifer tarandus*), horse (*Equus ferus*), and Norway lemming (*Lemmus lemmus*) have been found.

Upper Pleistocene floras and faunas

The Upper Pleistocene comprises an interglacial temperate stage sandwiched between two glacial stages. In Britain, the existence of a separate Wolstonian glacial stage is questioned by some researchers, and in Ireland no temperate stage deposits have yet been found.

Temperate stage

The single temperate stage (the traditional view) in the Upper Pleistocene is the Ipswichian, the type site for which is a lake site at Bobbitshole, near Ipswich in Suffolk. This was a comparatively short-lived temperate stage lasting about 11,000 years. The summary pollen diagram (see Figure 8.3) shows the general characteristics of floral changes during the Ipswichian. Not shown in this simple diagram is the rise in herbaceous pollen in the late mesocratic and telocratic phases. The reduction in the relative amount of tree pollen suggests the development of a more open aspect to the vegetation, and it is probable that some floodplain sites were extensively deforested perhaps by the grazing and trampling of large mammals. Beech (*Fagus sylvatica*) was apparently absent from Ipswichian forests, and was replaced by hornbeam, whose marked presence shortly after the thermal optimum is a characteristic feature of the Ipswichian temperate stage. It should be noted, however, that beech is a low pollen producer, and it is possible that it is under-represented in pollen records. Lime and alder are very scarce, as are *Abies* and *Picea*. In the continental Eemian interglacial (= Ipswichian) lime and *Picea* were much more important and it has been suggested that this might be evidence for the existence of a water barrier across the Channel at this time. The situation with regard to particular genera is, however, complex. For example, Huntley and Birks (1983) report the virtual absence of beech in the Eemian interglacial of Europe. Beech appears to be unable to invade closed forest, and its northward expansion from glacial refuges in the mountains of Italy, Yugoslavia, and Bulgaria, and coupled with its failure to enter Britain, might have been restricted by this factor alone.

The flora as a whole contains a number of species having a markedly southern range in Europe today, such as the Montpellier maple (*Acer monspessulanum*) and firethorn (*Pyracantha coccinea*) suggesting a July mean temperature for the interglacial optimum about 3 °C higher than at present.

The fauna of the Ipswichian interglacial is reasonably well known. The hippopotamus (*Hippopotamus amphibius*) is unknown from the Hoxnian interglacial, but is well recorded in and diagnostic of the Ipswichian.

According to Stuart (1982), the classical faunal assemblage also contains: spotted hyena (*Crocuta crocuta*), the straight-tusked elephant (*Palaeoloxodon antiquus*), fallow deer (*Dama dama*), and an extinct rhinoceros (*Dicerorhinus hemitoechus*). In the open forests of the telocratic phase, the woolly mammoth (*Mammuthus primigenius*) and the wild horse, or tarpan (*Equus ferus*), became more common. This horse, now found only in south-west Mongolia, is the ancestor of the domestic horse. Of the reptiles, the European pond tortoise (*Emys orbicularis*) was present until late in the interglacial cycle. Today, *Emys* is restricted to southern Europe where the mean July temperature exceeds 17–18 °C. Its presence in the Ipswichian suggests a warmer, more continental climate than at present (see Figure 8.5).

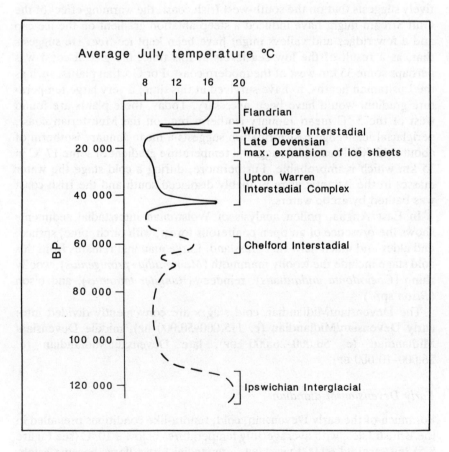

Figure 8.5 Variations in average Devensian July temperatures in lowland areas of southern and central British Isles. Chronology prior to c.50,000 BP uncertain due to limits of radiocarbon dating – dashed line. (Modified and redrawn from Coope 1977)

Glacial cold stages

During the Wolstonian/Munsterian cold stages, the British Isles was extensively glaciated. Very little of the North Sea basin remained ice-free and a joint British–Scandinavian ice-sheet deposited glacigenic deposits as far south as the Dogger Bank. Only southern England seems to have escaped glaciation.

In Ireland, Mitchell (1976) suggests that during the Munsterian, the whole island was covered by ice, but there is the possibility of higher ground in the south and west remaining ice-free as nunataks above the surrounding ice. Both fauna and flora were displaced by the Munsterian ice-cap and the question arises as to where the Gortian biota retreated. During this cold stage suitable refugia for the Gortian trees and shrubs probably did not exist north of the Pyrenees. Mitchell, however, tenta-tively suggests that on the south-west Irish coast, the warming effect of the Gulf Stream might have induced a steep ablation gradient on the ice cap and a few ridges and valleys might have been kept ice-free. He suggests that, as a result of the low sea levels at this time, the glacial coast was perhaps some 35 km west of the modern coast. For Gortian plants, such as the Lusitanian heaths, to have survived at this time, a very large tempera-ture gradient would have been necessary. Today, these plants are found west of the 5 °C mean January isotherm but, on the Munsterian coast, periglacial ice-wedge cast formation suggests a mean January isotherm of about −12 °C. These data suggest a temperature gradient of some 17 °C in 35 km which is improbable. Furthermore, during a cold stage the water masses in the Atlantic were probably displaced south and the Irish coast was bathed by arctic waters.

In East Anglia, pollen analysis of Wolstonian interstadial sediments shows the presence of an open coniferous forest, with birch, pine, spruce, and alder and much herb-rich grassland. Large mammal remains from this cold stage include the woolly mammoth (*Mammuthus primigenius*), woolly rhino (*Coelodonta antiquitatis*), reindeer (*Rangifer tarandus*), and bison (*Bison* spp.)

The Devensian/Midlandian cold stages are conveniently divided into: early Devensian/Midlandian (*c.* 115,000–50,000 BP); middle Devensian/Midlandian (*c.* 50,000–26,000 BP); late Devensian/Midlandian (*c.* 26,000–10,000 BP).

Early Devensian/Midlandian

For much of the early Devensian, cold, tundra-like conditions prevailed in the British Isles, with average July temperatures below +10 °C (see Figure 8.5) but, around 60,000 years BP, interstadial forest floras became estab-lished as temperatures rose. Remains of this *Chelford Interstadial* are known from Beetley and Wretton in Norfolk (West *et al.* 1974) and from the type site in Cheshire. At all three sites the vegetation comprised an open forest of birch, pine, and spruce. The pollen evidence also suggests the development of acid heath and bog communities as temperatures fell back toward the close of the interstadial episode. The insect fauna at Chelford comprised 100 species, including 3 now found in Scandinavia and the mountains of central Europe (Coope 1959). Although the Chelford vegetation is thought to be very similar to that found in northern Finland today, the presence of insects such as the pine weevil (*Hylobius abietis*) suggests a climate more resembling southern Finland, with a cool continental climate and a mean July temperature of +15 °C (see Figure 8.5). Among the mammal fossils recovered from deposits of this interstadial are an extinct bison (*Bison priscus*) and reindeer (*Rangifer tarandus*) whose grazing herds would have been common, especially during herbaceous phases (Stuart 1982).

Middle Devensian/Midlandian

Between about 45,000 and 25,000 years ago a cool, and probably variable, continental climatic regime with cold winters became established and interstadial floras and faunas flourished both in Britain and Ireland. This *Upton Warren Interstadial Complex*, named after the type site at Upton Warren in Worcestershire, was a period of tundra devoid of trees, with the mean July temperature probably just below +10 °C (see Figure 8.5). During the short, warm, thermal maximum of the Upton Warren Interstadial Complex, at about 43,000 BP a rich thermophilous insect fauna of more than 300 species was present. Mean July temperatures are thought to have been about +18 °C. (Coope, Shotton, and Strachan 1961). The absence of trees at this time cannot have been due to the lack of warmth, and probably reflects their inability to migrate into the British Isles from southern refuges, in the thousand years or so that the amelioration lasted.

The continentality of the climate at this time was demonstrated by Bell (1970), in her study of macroscopic plant remains at Earith near Huntingdon. The presence of plants such as slender naiad (*Najas flexilis*) and gipsywort (*Lycopus europeus*) indicate a July temperature of at least 16 °C. At the same site, the development of penecontemporaneous periglacial ice-wedges and involutions indicates a mean annual temperature of not more than −6 °C, and winter temperatures, therefore, considerably lower. By about 40,000 BP, tundra conditions had returned, and the southern element of the beetle fauna (e.g. *Calosoma reticulatum*, *Hister funestus*, *Rynchites pubescens*) had died out and boreal species took their place. The presence of the dung beetle (*Aphodius holdereri*) in this interstadial is interesting because today it is restricted to the high plateau of Tibet above 3,000 m (Coope 1979).

Bison and reindeer dominate the vertebrate fossils of the Upton Warren

Interstadial Complex, but remains are also present in some quantity of mammoths (*Mammuthus primigenius*), woolly rhinoceros (*Coelodonta antiquitatis*), and horse (*Equus ferus*).

In Ireland, two middle Midlandian interstadial sites have been described. At Hollymount, near Linaskea in County Fermanagh organic silts, with a radiocarbon age of more than 41,500 BP, beneath the till of a drumlin, contain vegetable debris derived from a tundra landscape. A few miles away, at Derryvree, organic silts and fine sands beneath a drumlin have been dated to 30,500 BP. They contain a rich flora of seeds and mosses and the pollen of grasses and sedges (Edwards and Warren 1985).

Little is known about the mammalian fauna in Ireland at this time, although a cave deposit at Castlepook, County Cork, has produced a substantial fauna including:

Alopex lagopus (arctic fox);
Canis lupus (wolf);
Crocuta crocuta (spotted hyena);
Dicrostonyx torquatus (Greenland lemming);
Mammuthus primigenius (woolly mammoth);
Equus caballus (horse);
Lemmus lemmus (Norwegian lemming);
Lepus timidus (mountain hare);
Magaceros giganteus (Irish giant deer);
Rangifer tarandus (reindeer);
Ursus arctos (brown bear).

A mammoth bone from this site has been radiocarbon dated to 33,500 BP, and the fauna probably existed during the Derryvree Interstadial. Mitchell (1976) suggests that the presence of mammoths and giant deer indicate a rather lush vegetation as bare tundra would probably not have provided enough food for such large herbivores. The giant deer was, of course, not 'giant' and although it had a very large antler span (up to about 3.5 m), its body size was smaller than that of the modern North American moose.

Late Devensian/Midlandian glacial periods

During the late Devensian/late Midlandian, most of the British Isles was inundated by thick ice sheets (see Figure 8.6). Precise details of the amount of glacio-eustatic lowering of sea level at this time are not known, but it was probably of the order of 100 m and, where not covered by ice, exposed parts of the continental shelf were subject to the rigours of a periglacial environment. These marginal areas, part land, part exposed sea-bed, must have acted as '*gross refugia*' for those plants and animals that could have withstood the climate. Mitchell suggests that some northern species recorded from the middle Midlandian interstadial, and which still maintain

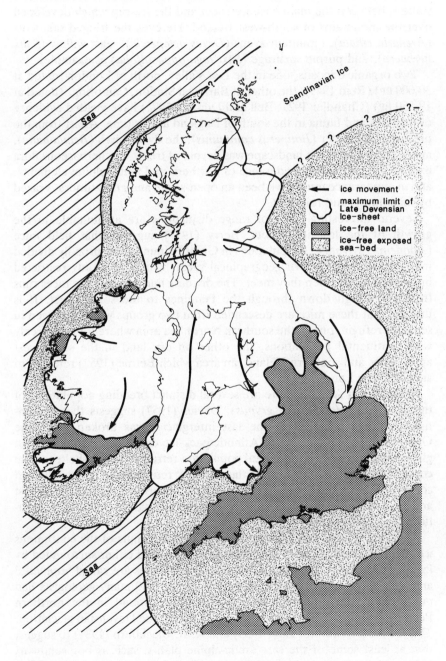

Figure 8.6 Limits of the Devensian ice sheet. (Based partly on Bowen *et al*. 1986)

a foot hold in western Ireland might have survived at more southerly stations between the main Irish ice sheet and the ice-cap which developed over the mountains of south-west Ireland. He cites the fringed sandwort (*Arenaria ciliata*), mountain avens (*Dryas octapetala*), least willow (*Salix herbacea*), and purple saxifrage (*Saxifraga oppositifolia*) as examples.

Two organic deposits, one in the Lea Valley, north of London (dated at 30,000 BP) (Reid 1949), the other at Barnwell Station, Cambridge (dated at 19,550 BP) (Chandler 1921; Bell and Dickson 1971) have yielded evidence of the flora and fauna in the south of England at this time. Apart from the presence of Juniper (*Juniperus communis*), the dwarf birch (*Betula nana*), and arctic willows, the landscape was devoid of trees. The herbaceous flora exhibited a strong preponderance of northern and arctic-alpine elements, and in appearance must have been an open-sedge and grass-tundra, grazed by mammoths and reindeer.

Several interesting cases of range disjunctions are attributable to the growth of the last ice sheets. Berry (1977) suggests that field mice (*Apodemus sylvaticus*) populations in Great Britain can be broadly divided into western and eastern geographical races, which have only restricted breeding success when they meet. The dividing line between the races runs from Edinburgh down through the Pennines to the south coast. It is thought that these mice are descended from two groups which diverged in separate refugia, one in the southern North Sea area where it mixed freely with continental populations, the other on dry land in the south-west approaches and southern Ireland, an area which Beirne (1952) referred to as the 'Celtic Land' refuge.

In the North Atlantic there are several disjunct breeding populations of the grey seal (*Halichoerus grypus*). Davies (1957) suggests that its continuous range achieved in the last interglacial was broken when the Greenland ice cap and North Atlantic pack-ice developed during the last glacial maximum. Part of the seal population retreated to the shelter of the coast of the United States, while the rest took refuge along the unglaciated coasts of France and Spain. With the return to warmer conditions, the east and west Atlantic populations have remained separate breeding populations (see Figure 8.7).

A second type of refugium occurred within the limits of the continuous ice sheet, and consisted of a restricted, ice-free area either surrounded by ice, or bordered by the sea on one side. These types are called '*nunataks*' and '*coastal refugia*', respectively. Until about 1950 many geologists had put forward the argument that much of the ground in the northern Pennines above *c*. 670 m OD was ice-free during the Devensian glaciation. This led A. J. Wilmott (1930, 1935) and K. B. Blackburn (1931) to suggest that at least some of the rare arctic-alpine plants, such as bog sandwort (*Minuartia stricta*) growing at the head of Teesdale, survived the glaciation on local nunataks around the head of the dale. Nowadays it is thought

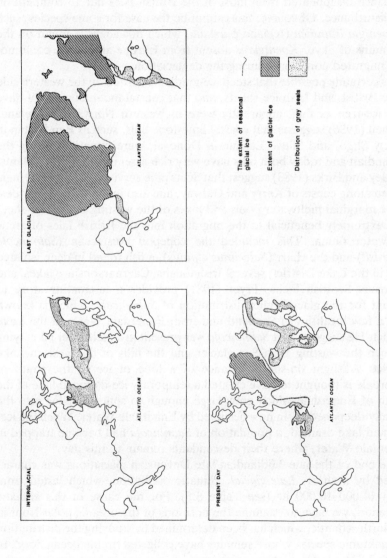

Figure 8.7 The glacial origin of the range disjunction of the grey seal (*Halichoerus grypus*). (Modified and redrawn from Davies 1957)

unlikely that ice-free areas existed in the northern Pennines, and pollen analyses have shown that many rare plants were actually present in the area, and over much of Britain, as components of a widespread Late-glacial flora that rapidly spread northward to colonize sites when the ice melted (Clapham 1978). This once-widespread, light-demanding flora has long since disappeared from most of the British Isles due to competition and disturbance. Of course, this cannot be the case for some species, such as *Koenigia islandica* (Iceland purslane) which nowadays is found on the mountains of Skye. *Koenigia* is absent from Europe and hence could not have migrated northward during the deglaciation.

It is certainly possible that steep ice-gradients existed on the western side of the British and Irish ice-sheets, and that coastal mountains might have been ice-free as they apparently were in western Norway. In Ireland, Mitchell (1986) suggests that coastal limestone hills, such as Ben Bulbin in County Sligo, and Slieve League in Donegal, were nunataks during the Midlandian and today both hills have very rich assemblages of rare plants. Huntley and Birks (1983) suggest that Scots pine survived on glacial refugia located along coasts of Kerry and Galway, and also in the Outer Hebrides.

The marginal meltwater pools and lakes of the wasting British ice-sheet were extremely beneficial to the migration into the British Isles of arctic freshwater fauna. This included the copepod crustacean (*Salmincola edwardsii*) and the charr (*Salvelinus alpinus*), a fish found in deep isolated lakes in the Lake District, several Irish loughs, Caernarvonshire lakes, and numerous Scottish lochs. Fryer (1981) provides a fascinating story to account for the glacial relict distribution of *Salmincola*, which is known from a few localities in Scotland and from Ennerdale Water in the Lake District. He suggests that *Salmincola* was present in a meltwater lake, lying between the wasting Irish Sea glacier and the hills of the western Lake District. A slight on-shore advance of a lobe of ice to the south of Ennerdale is thought to have created a temporary ice-dammed lake at the mouth of Ennerdale which grew large enough to spill up valley into the glacially-deepened basin now occupied by Ennerdale Water. When the ice-dammed lake drained, a population of *Salmincola* had become trapped in Ennerdale Water, where their descendants remain to this day.

The end of the late Midlandian/late Devensian glaciations was charac-terized by a short, *Late-glacial*, climatic oscillation, which lasted from about 14,000–10,000 BP (see Table 8.5). For the cause of this climatic oscillation, we have to examine the positions of the oceanic polar front in the North Atlantic, which has been determined by studying the distribution of planktonic species whose remains have collected on the ocean floor. In Figure 8.8 the position of the polar front can be seen to have flipped north–south about the British Isles, directly inducing major changes in the thermal characteristics of the waters bathing the continental shelf, and indirectly affecting the whole climatic regime of oceanic western Europe.

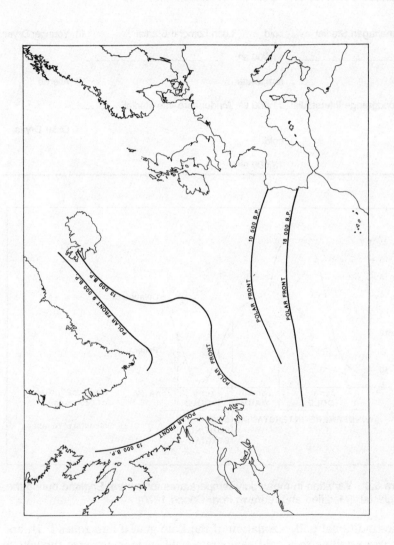

Figure 8.8 Position of the oceanic polar front in the north Atlantic between 18,000 BP and 9,000 BP. (Modified and redrawn from Ruddiman and McIntyre 1981)

Table 8.5 Subdivisions of the Late-glacial

Ireland		Britain	Pollen zone	Europe
Littletonian Post-Glacial ----------- 10 000 BP		Flandrian --------------------		
Nahanagan Stadial	cold	Loch Lomond Stadial		III Younger Dryas
-------------------- 11,000 BP		----------------------------		
	temperate			II Allerød
Woodgrange Interstadial 11,800 BP		Windermere Interstadial		--------
				I Older Dryas
	cold			
13,000 BP --				

(Left margin vertical labels: LATE — GLACIAL)

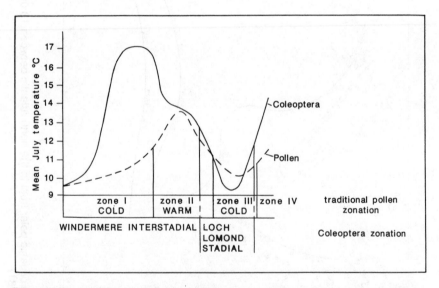

Figure 8.9 Variation in mean July temperatures for central England during the Late-glacial. (Modified and redrawn from Coope 1970)

The traditional pollen zonation of the Late glacial into zones I, II, and III, corresponding to a cold–temperate–cold cycle is now not thought to represent the actual environmental changes, because the pollen evidence reflects the rather slow response of the flora to temperature changes. Coope (1970) inferred climatic changes based on the evidence of fossil thermophilous ground beetles (order Coleoptera) which, he argued, could migrate much more quickly than plants. Coope's findings are illustrated in

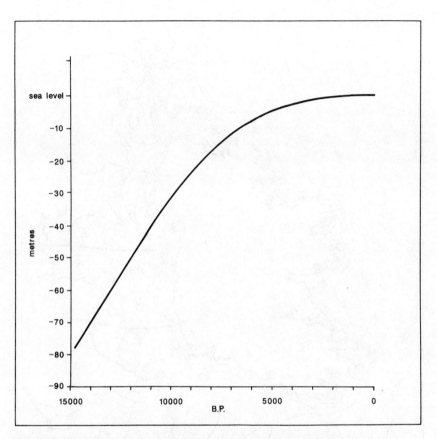

Figure 8.10 Simplified Late-glacial and Flandrian sea-level curve

Figure 8.9, where it can be seen that the thermal maximum of the Late glacial, as indicated by beetle evidence, was not only earlier but also warmer than the regime suggested by pollen analysis. By about 13,000 BP mean July temperatures were as warm as, if not warmer than, present conditions, but by 10,500 BP the British Isles was once again in the grip of arctic conditions, and small glaciers developed in the highest mountains in Scotland and the Lake District.

The flora of the Windermere/Woodgrange Interstadial was extremely species-rich, although dominated by northern and alpine elements. Among the characteristic genera of the open tundra habitats were: *Rumex* (docks); *Artemisia* (mugworts); *Armeria* (thrift); *Plantago* (plantains); *Galium* (bedstraws); *Epilobium* (willow herbs); *Polygonum* (bindweeds); *Succisa* (devil's bit scabious); and *Valeriana* (valerians) (Pennington 1969). Birch, arctic willow, and juniper were widespread at the peak of the interstadial (as determined from pollen) and pine was well established in the south of

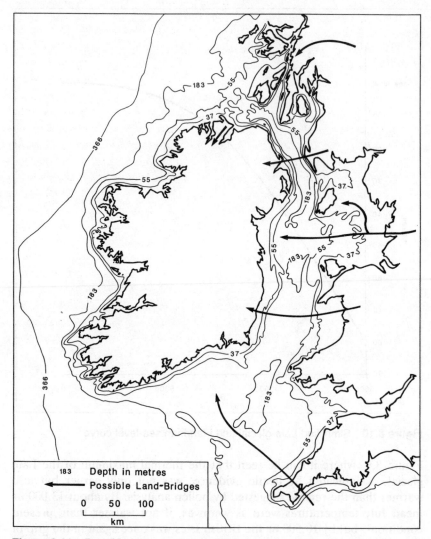

Figure 8.11 Possible Late-glacial land-bridge routes into Ireland. (Redrawn from Devoy 1985)

England. In the open, herb-rich tundra meadows of Ireland, the Irish giant deer (*Megaceros giganteus*) reached its zenith only to become extinct in the succeeding cold spell, the Nahanagan Stadial.

The expansion of the Irish flora during the Woodgrange Interstadial was partly due to a massive immigration of new plants and animals and, according to Mitchell (1986), this in itself was evidence of land connections between Britain and Ireland at this time. The precise position of sea level is difficult to determine, because we must not only take into account global sea-level (eustatic) changes, but also local isostatic movements due to glacier loading and unloading. Figure 8.10 shows a simplified eustatic sea-level curve which suggests that sea level was about −95 m at the beginning of the Late-glacial.

Five possible land-bridge connections between Ireland and Britain have been suggested as post-glacial invasion routes (Devoy 1985). These are shown in Figure 8.11 together with a simplified bathymetry. For much of its length the Irish Sea is floored by a more or less linear trough reaching no higher than about −90 m OD today. This trough is probably a Miocene structure, and would have provided the sea with a ready access to the area soon after glaciation. Devoy provides a detailed analysis of each route, taking into account what is known of the isostatic rebound in each area. He suggests that the Celtic Sea route from south-west England to southern Ireland is improbable, and the Irish Sea routes all appear somewhat illusory. However, between the Inishowen Peninsula in Northern Ireland and Islay in Scotland, the sea is relatively shallow and between 11,400 and 10,200 BP a string of islands and short-lived peninsulas, perhaps some 2 to 6 km wide could have acted as land-bridges. It has also been suggested that this route was exposed prior to about 12,000 BP but this is evidently not supported by the available data on sea-level changes in the area. However, a slightly earlier date is necessary to accommodate the influx of biota into Ireland during the Woodgrange Interstadial.

In spite of Devoy's evidence on the bathymetry and positions of relative sea levels, the fact remains that animals and plants did manage to colonize Ireland. For example, oak trees, red deer and the wild boar were present in Ireland some 8,500 years ago. Mitchell, a great advocate of land-bridges, prefers a scenario of organized oakwoods, replete with forest animals, advancing across a land-bridge, rather than to imagine the occasional acorn floating across the sea or being carried in the crop of a bird while groups of pigs and deer were swimming across tidal channels. Perhaps Mitchell's ideas of morainic ridges between the Lleyn Pensinsula in Wales and Wicklow on the Irish coast, and also between Anglesey and Drogheda, are not too far-fetched (Mitchell 1963; Preece, Coxon, and Robinson 1986). The problem here is that although undoubted ridges do exist, much has been removed by subsequent erosion and we really have no idea how large they were originally and for how long they might have acted as invasion routes.

On the other hand, it is difficult to see why typical woodland mammals – such as the mole, weasel, and fallow deer – never made the crossing. Indeed, the absence in Ireland of fallow deer and elk – which were well established in England in the Pre-Boreal – suggests that any land-bridge was flooded before the English forest fauna could invade.

Land connections between Europe and England in the Late-glacial are

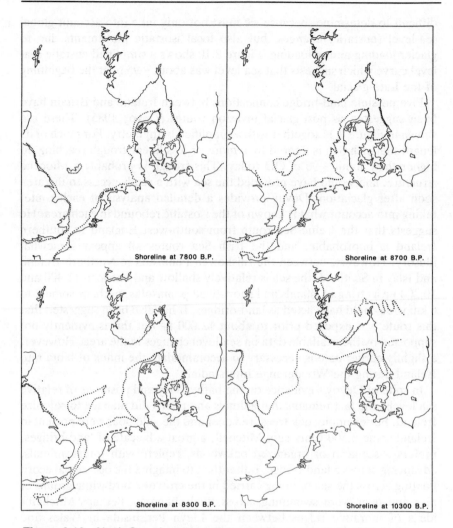

Shoreline at 7800 B.P.

Shoreline at 8700 B.P.

Shoreline at 8300 B.P.

Shoreline at 10300 B.P.

Figure 8.12 North Sea shorelines during the Devensian Late-glacial and Flandrian stages. (Redrawn from Jelgersma 1979)

less problematical. Jelgersma (1979) has reconstructed Late-glacial and early Flandrian shorelines in the North Sea and eastern English Channel (see Figure 8.12). For much of this time, large areas of the southern North Sea basin were land, across which plants and animals could spread rapidly into the British Isles. We should note that the land connection was not simply a southern one into northern France and the Low Countries, but also linked the British Isles with the north European Plain, whose continental biotic elements were some of the first invaders into eastern England.

The mixing of British and European faunas and floras took place not only on land but also in fresh water, as rivers from the continent and eastern England merged before draining through the English Channel (Beaufort 1951). This palaeogeography probably explains the distribution of fishes such as the barbel (*Barbus barbus*), the eelpout (*Zoarces viviparus*), and the white bream (*Blicca bjaernka*), which occur naturally in the rivers of eastern England, and are also found in the Rhine.

Littletonian/Flandrian Stages

The climatic recession which produced Britain's last glaciers came rapidly to an end about 10,000 BP, as temperatures rose during the end of the protocratic phase of the present interglacial. The last protocratic phase includes both the Late-glacial and the earliest part of the Flandrian/Littletonian stage from 10,000 to 9,000 BP, in the scheme described by Birks (1986). A general picture of the environmental changes which took place in Britain over the last 10,000 years is shown in Figure 8.13. Apart from one or two details which are mentioned below, the development of the Irish environment was similar to that of Britain.

By the end of the protocratic phase (pollen zone IV), birch and then pine were widespread, their rapid dispersal aided no doubt by their wind-borne propagules. The presence of some thermophilous species, such as the great sedge (*Cladium mariscus*) indicates that summer temperatures might have been as high as they are today. More warmth-demanding trees, such as the oak and elm, could have survived in the British Isles at this time, but their dispersal northward from their refuges in southern and south-eastern Europe was slow, reflecting the disequilibrium between climate and tree distributions at this time (Birks 1986).

Many large mammal species present in the Late-glacial did not survive into the Flandrian/Littletonian stage. The mammoth, woolly rhinoceros, bison, giant deer, and reindeer died out in the British Isles. The cause of this extinction is not known for sure. In Ireland it could not have been due to man since he did not arrive until about 9,000 BP during the Mesolithic. In Britain it is just conceivable, though improbable, that small populations of upper Palaeolithic hunters using primitive hunting techniques wiped out these larger mammals.

According to Birks (1986) the early temperate *mesocratic phase* lasted from about 9,000 to 5,000 BP (pollen zones V–VIIa) and was characterized by the dispersal of temperate tree species and the development of brown-earth forest soils. During the so-called *climatic optimum* – from about 8,000 to 5,000 BP, summer temperatures were probably about 2–3 °C higher than they are at present. Tree species did not disperse as a vegetation assemblage. Each species underwent a phase of dispersal into the British Isles and, as conditions became favourable, they rapidly

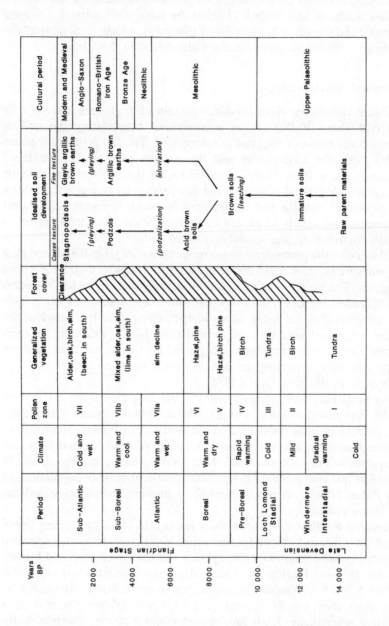

Figure 8.13 Relationships between climate, pollen zones, vegetation, and soils in the Devensian Late-glacial and Flandrian stages. (Modified and redrawn from Bridges and Davidson 1982)

expanded their range (see Table 8.6). At least some of this range expansion must have been aided by animals otherwise it is difficult to understand how some mesocratic trees could have achieved such high annual dispersion rates. The slow dispersal rate of *Tilia*, might be one reason why it failed to colonize Ireland before rising sea levels had submerged any possible land-bridge connection. Dispersal into Britain from Europe was effectively stopped about 8,300 BP, when the sea finally flooded through the Straits of Dover (see Figure 8.12).

At their peak, the mesocratic forests covered all but the highest and most inhospitable parts of the British Isles. Over much of England and Wales, the forest climax was a mixed oak woodland, with elm and lime. In south-eastern England, beech and hornbeam were naturally dominant or co-dominant on suitable soils. The forests of Scotland were dominated by pine apart from the far north-east where birch was the climatic dominant. In Ireland oak and elm flourished, and had displaced pine on the poorer soils. Godwin (1975) describes the climax forests as a polyclimax rather than a monoclimax in which local conditions of altitude, aspect, drainage, and geology could determine variation from broad climatic climax.

Table 8.6 Main expansion periods of mesocratic trees

Genus/species	Expansion period (BP)	Average dispersion rates (km yr^{-1})
Corylus avellana (hazel)	9,500–9,000	1.5
Ulmus (elm)	9,000–8,000	1.0
Quercus (oak)	8,500–7,500	0.5
Tilia (lime)	8,000–6,000	0.2
Fraxinus excelsior (ash)	6,000–5,000	0.2

Sources: Birks 1986; Huntley and Birks 1983

Soil deterioration, a cooler, moister climate, and the arrival of Neolithic man, mark the beginning of the *oligocratic* or late temperate phase of the Flandrian/Littletonian stage at about 5,000 BP. In the oceanic climate of the British Isles, it is doubtful if the soil fertility under a dense mesocratic forest could have been maintained permanently and the gradual podsolization and impoverishment of the soil was a key mechanism bringing about important vegetation changes. In the uplands of the British Isles podsolization was accompanied by descending tree-lines, and the expansion of acid grasslands and *Calluna*-dominated heaths, and, where drainage was impeded, the development of blanket bog communities. Godwin suggests that the progress of soil degeneration was much quicker on the igneous rocks of the highlands than the younger sedimentary rocks of the lowlands: correspondingly, the teleocratic phase of the interglacial cycle would have been reached earlier in the north, thus emphasizing the latitudinal gradient.

The impacts of prehistoric man

It is difficult to separate the effects of edaphic and climatic changes in the early oligocratic, from those of man. It was at this time, that Neolithic farmers began what was to be a more or less continuous series of impacts on the flora and fauna of the British Isles that have continued up to the present day. A detailed account of the influence of prehistoric man on the British vegetation is beyond the scope of this chapter and it has been a central theme for much palynological research. The interested reader will find a very full account in *The Environment in British Prehistory* (Simmons and Tooley 1981).

Pollen diagrams from all over the British Isles and north-west Europe show a more or less synchronous, and rapid, decline in the amount of elm pollen at about 5,000 BP (3,000 BC). Birks (1986) notes the following important features of the so-called *elm decline*: (i) its synchroneity throughout the geographical range of *Ulmus*, even in areas not known to be colonized by Neolithic peoples (e.g. northern Scotland); (ii) the rapid decline in pollen percentages over a period of 50 to 100 years, suggesting that the decline did not take place over several generations; (iii) its consistency from site to site with little evidence that other thermophilous trees were similarly affected; (iv) elm populations rarely re-established their former size.

Iversen (1941) suggested that the elm decline was due to climatic change. In particular, he thought that the climate became more continental with the result that the colder winters would adversely affect the production of flowers in the elm.

A theory put forward by the Danish botanist Troels-Smith (1960) suggests that the decline was caused as a result of local forest clearance (landnam clearances) by Neolithic farmers who used slash and burn techniques, and perhaps ring-barking, to make small garden plots in which they grew cereals and sheltered their livestock. The introduced cereals included emmer wheat (*Triticum dicoccum*), naked six-row barley (*Hordeum vulgare,* var. *nudum*), and hulled six-row barley (*Hordeum vulgare*) which is still grown in Britain today. Elm and hazel leaves were also collected from winter fodder. As soil fertility declined, the small family units would have had to move on leaving their clearing to be colonized by invasive, shade-tolerant trees such as ash and hazel and later, limes, and oak.

A third theory suggests that the decline of elm was due to a widespread pathogenic attack not dissimilar to the Dutch elm disease of recent times (Rackham 1980). Very probably, the decline of the elm was due to both man and disease. Certainly, disease-damaged trees would have been that much easier to clear and, indeed, man may have helped the spread of the disease.

Neolithic farmers introduced a number of animals into the British Isles, including sheep, goats, pigs, and cows. Not only would the grazing and browsing activities of these domestic animals have been in competition with indigenous species, such as deer, but they would also have helped suppress forest regrowth and assist in maintaining the newly opened grassy glades.

During the Sub-Boreal, about 4,000 BP (2,000 BC) the British Isles began to be colonized by Bronze Age settlers. In some areas, Bronze-Age peoples continued the tradition of temporary woodland clearances and practised a small-scale arable and pastoral economy. Elsewhere, in the Cumberland lowlands, the chalk hills of southern England, and Ireland, clearances for agriculture were much more extensive, and more-or-less permanent deforestation had set in. Forest clearance, and the creation of heath and bog, was particularly widespread on the upland fringes of Dartmoor, the Pennines, and the North York Moors (Turner 1970).

One notable effect of Bronze Age man on the fauna was the apparent extinction of the auroch (*Bos primigenius*) which had been present in the British fauna from at least Hoxnian times. This large, ox-like bovine, was probably hunted both for its meat and horn, but it would also have been a nuisance around the unfenced cereal crops.

By the end of the Sub-Boreal (c. 2,500 BP), the climate had deteriorated quite markedly, water tables rose, soils became wetter, and there was active growth of peat. These environmental changes mark the beginning of the Sub-Atlantic period, a period which also saw the entry of Iron Age man into the British Isles. The use of iron had a profound effect on the vegetation. Not only were iron axes much more effective tools for tree felling than those made of bronze, but also their manufacture required wood which was a further reason for forest clearance. Equally important was the development of the iron plough which allowed soils to be converted for arable crops grown in small (0.2–0.6 ha), rectangular, so-called 'Celtic' fields in areas which previously had been uncultivated because of difficult conditions.

In southern England, the clearances which had began in the Bronze Age continued, and by the time of the Roman invasion in AD 43 the landscape was essentially open and dominated by a well-established, sheep-farming economy (Evans 1975). Elsewhere in Britain the lowland forests, which were more or less intact during the early Iron Age, were progressively opened up. Only in north-west England and Scotland did a forested landscape remain.

In Ireland the picture at this period is unclear. Mitchell (1976) suggests that farming was disrupted over wide areas of the central lowlands from about 300 BC. Pollen indicative of arable land virtually disappears, and an increase in hazel and elm suggests the widespread development of secondary forests. This was the time of the Pagan Iron Age, an obscure period of Irish history. The disruption came to an end about AD 300, when the pollen evidence shows a devastating clearance of ash and elm possibly encouraged by the introduction of ploughs with iron coulters, a downward-projecting metal knife, which was much more efficient than the wooden or stone ard at cutting through matted roots.

By the beginning of the historical period, the general pattern of the flora and fauna was established. Perhaps the most striking fact is the paucity of the flora and fauna of Ireland, as compared with that of Britain, and the paucity of the British flora and fauna as compared with that of the Continent. One recent estimate for numbers of flowering-plant species is given by Webb (1983). After taking into account naturalized aliens and microspecies in genera such as *Rubus*, *Hieracium*, and *Taraxacum* the following figures are arrived at:

	Number of Species
France	3,500
Britain	1,172
Belgium	1,140
Denmark	1,030
Ireland	815

The poverty of the Irish flora is even more striking, if it is realized that Ireland is nearly three times the size of Belgium and has considerably more habitat diversity. There are similar shortfalls in other taxa:

	Ireland	Britain
mammals	28	55
birds	354	456
reptiles	1	4
amphibians	2	6

From Roman times up to the present day, man has continued to have enormous impacts on the flora and fauna, whether they be by the introduction of alien species, by the further destruction and contamination of natural habitats, or by the consumption of land for urban areas and all their necessary facilities. These changes will be considered in the next two chapters.

Chapter nine

The thirst for land and loss of habitats

With few notable exceptions, the bulk of the native terrestrial flora and fauna of the British Isles has suffered range contraction and fragmentation. This loss has, in large measure, been due to habitat loss and to the impacts of an increasingly mechanized agricultural system over the last 2,000 years, and particularly over the last 100 years or so. Changes in the distribution of British butterflies, a well-known group, illustrate this point well. Of the sixty-two species considered in the *Atlas of Butterflies in Britain and Ireland* (Heath, Pollard, and Thomas 1984), twenty-four have suffered major range contraction, and four are no longer found in the British Isles. Such profound changes are not at all unexpected, given the pace of land-use change in many parts of the British Isles. For example, the Council for the Protection of Rural England has recently estimated that in the last 40 years an unspoilt area the size of Berkshire, Bedfordshire, Buckinghamshire, and Oxfordshire has disappeared under concrete; 95 per cent of our flower-rich meadows have gone under the plough; 60 per cent of our lowland heaths been lost; 50 per cent of our ancient woodlands cut down; 50 per cent of our lowland fens drained; and every year some 7,000 km of hedgerow disappear.

Today we live increasingly in a world of agribusiness, where accountants often appear to wield more influence on the survival of a species than biogeographers and ecologists. Short-term economic gains won by damaging activities – such as the draining of flood meadows and the removal of hedges – are having profound effects on the supply of habitats for our flora and fauna. Although this chapter is concerned with the terrestrial biota it is abundantly clear that marine species, too, have been seriously affected by increased commercial exploitation. The over-fishing of the herring (*Clupea harengus*) in the North Sea is well known. More recently, the skate (*Raja batis*) seems to have suffered an equally disastrous population decline and is hardly to be found in the Irish Sea.

There are some taxa, particularly invasive aliens, such as the collared dove (*Streptopelia decaoto*), which have actually been remarkably successful in expanding their range within the British Isles, but these are very much the exception. Usher (1986a) gives an account of invasive species in the context of wildlife conservation. One alien plant with the potential to displace, and therefore reduce, the range of native species is the rhododendron (*Rhododendron ponticum*) which was introduced into the British Isles in 1763 as an ornamental shrub. In south-west Ireland, rhododendron has become established in the Killarney oakwoods, whose mild oceanic climate supports many mosses, liverworts, and lichens. As an understorey species, rhododendron seems to have a number of important ecological effects. First, its dense foliage reduces the amount of light falling on the forest floor and this has had serious consequences for the epiphytic bryophyte and lichen flora, with species reductions of up to 75 per cent being recorded. Second, its unpalatability to grazing animals means that it is a lot less susceptible to overgrazing than native species such as holly (*Ilex aquifolium*) which it is apparently displacing. In the case of the Killarney oakwoods, the overgrazing is due to the introduced sika deer (*Cervus nippon*).

Of course, the often dramatic range contraction shown by some native species is not entirely a matter of habitat modification, destruction, or ecosystem simplification. In recent times there have also been significant, though small, climatic fluctuations, such as the so-called *Little Ice Age* (peaking at around 1750) which ought to be borne in mind (Ford 1982).

Until the development of biogeographical recording schemes much of the spatial detail of the changes in the flora and fauna remains vague, but a useful general picture of these early influences can be had from archaeological and historical sources (Sheail 1980).

The historical period

According to Rackham (1976), the first twelve centuries after the birth of Christ were undoubtedly the most important period in the formation of the British countryside, and much of the modern landscape was already recognizable. Most of our villages came into being and the proportions of farmland, moorland, and woodland were not enormously different from what they are now. Villages generated the need for permanent trackways and some, such as the Icknield Way and the Ridgeway, exist as such today.

Many trackways are lined by ancient hedges, which can be dated approximately using regression estimates of the invasion rates of woody species calibrated by documentary evidence. Roughly speaking, a 30 m stretch of hedge gains one new species for each century it has stood. Numerical refinements of this method developed by Pollard *et al.* (1974) give the following predictive equation:

$$\text{age} = (100 \times \text{number of woody species}) + 30$$

When this equation is applied to hedges in many parts of southern Britain, it predicts that they date back more than a thousand years.

With the invasion of England by the Romans came the introduction of the heavy mould-board plough, which inverts the cut sod and made it possible to open up some of the heavier soils in the clay vales of southern England. Pollen analyses show that, during this time, the landscape was dominated by an arable farming system, and, indeed, one of the main reasons for the Roman invasion was to tap the cereal-growing potential of much of lowland England and Wales. The movement of troops and the import of grain supplies was accompanied by accidental introduction of several plant species, especially those weeds which grow alongside cereal crops such as the corn marigold (*Chrysanthemum segetum*), the corncockle (*Agrostemma githago*), and the sow-thistles (*Sonchus oleraceus* and *S. asper*). All of these would have been actively spread as organized communications built up along the Roman road system.

The Romans also introduced a number of economically useful crops some of which are now established ruderals, but others are only found cultivated, or as garden escapes (see Table 9.1). A number of insect pests were also accidentally brought in alongside grain imports, such as the saw-toothed grain beetle (*Oryzaephilus surinamensis*) and the grain weevil (*Sitophilus granarius*).

Table 9.1 Economically useful plants introduced by the Romans

Belladonna	(*Atropa belladonna*)	Pea	(*Pisum sativum*)
Coriander	(*Coriandrum sativum*)	Plum	(*Prunus domestica*)
Dill	(*Peucedanum graveolens*)	Radish	(*Raphanus raphanistrum*)
Fennel	(*Foeniculum vulgare*)	Vine	(*Vitis vinifera*)
Medlar	(*Mespilus germanica*)	Walnut	(*Juglans regia*)

Source: Godwin (1956)

The conversion of land from forest to farmland meant that only in the north and west of Britain, and in the west of Ireland (Ireland was never invaded by the Romans), did much primary woodland survive. The term primary woodland is used here to denote a site which has been continuously wooded, as a climax ecosystem, since its Flandrian development. An evocative term for such woodland used frequently by Oliver Rackham, a noted authority on forest history, is *wildwood*.

By late Romano-British times, about AD 300, cereal cultivation was practised not only in lowland Britain but also in the uplands. In the Lake District, for example, several pollen diagrams record deforestation and cereal production as high as 225 m OD, higher than at any time since. In Ireland, too, there were the devastating clearances of elm and ash at about this time, and the land was made free for arable crops. These Irish clearances, together with other farming impacts in the historical period,

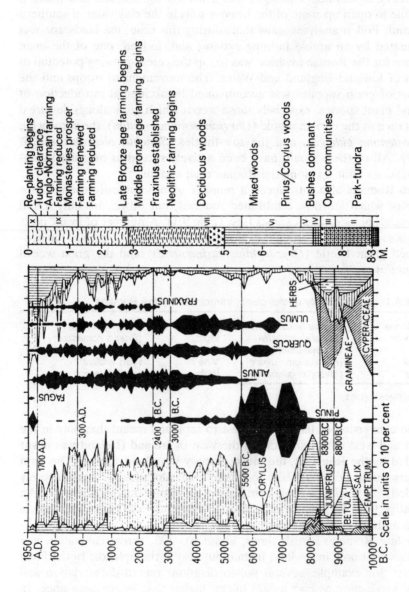

Figure 9.1 Pollen diagram extending from the Late-glacial to the present – Littleton Bog, Co. Tipperary. (From Mitchell 1965)

are clearly seen in Mitchell's pollen diagram from Littleton Bog, in County Tipperary (see Figure 9.1).

Throughout the Anglo-Saxon period, further inroads into the remaining wildwood occurred, and ox-teams coupled to mould-board ploughs made light work of soils which had previously been left unworked. Several pollen diagrams add detail to this general picture. For example, Godwin's pollen diagram from Old Buckenham Mere, Norfolk, shows, in addition to woodland clearance, a reduction in the pollen of pastoral weeds such as *Plantago* (plantains) and *Rumex* (docks) together with an increase in the pollen of Compositae, Cruciferae, and Chenopodiaceae species (Godwin 1967). These pollen taxa reflect a change of emphasis from pastoral farming to arable.

It is thought that the mould-board plough reached Ireland about AD 600. Mitchell suggests that the turning of the soil-sod by this plough buried some weeds, but that others, such as mugwort (*Artemisia vulgaris*), put down deep roots and survived. To judge by their increased presence in pollen diagrams, such plants became a widespread nuisance in arable areas.

Post-Domesday

By the Middle Ages, documentary evidence from sources such as the great Domesday survey of 1086, depict a late stage in the destruction of the primary forest in England. Rackham (1976) suggests that the last approximation to anything like wildwood may have been around AD 1150 in the Forest of Dean – which, within 100 years, was to be extensively coppiced and grazed. In Ireland, the Anglo-Norman influence was profound, and a manorial farming system became widespread on all but the peat and poorly-drained gley soils of the west and north-west of the country.

The growth of the monastic movement at this time saw a huge increase in sheep farming. These sheep not only destroyed saplings, but effectively stopped woodland regeneration through their selective grazing. The plagioclimax vegetation resulting from upland forest destruction varied according to soil conditions and subsequent land use. Where grazing was heavy, grasslands prevailed, with species such as sheep's fescue (*Festuca ovina*), brown bent grass (*Agrostis canina*) and wavy hair grass (*Deschampsia flexuosa*) dominant on drier skeletal soils, and brown earths, and mat-grass (*Nardus stricta*), purple moor-grass (*Molinia caerulea*) and the heath rush (*Juncus squarrosus*) on wetter gleys and podsols. Where grazing was lighter, a dwarf-shrub heath developed – with ling (*Calluna vulgaris*), bell heather (*Erica cinerea*), and bilberry (*Vaccinium myrtillus*) (Ratcliffe 1984). Although the Normans were responsible for the introduction of the rabbit (*Oryctolagus cuniculus*), they were mostly kept in warrens and were not ecologically important in the wild until the nineteenth century.

The effect of all these changes on the woodland fauna and flora is difficult to judge, but it seems that most of the larger mammals managed to survive, even though their populations must have been greatly reduced. In 1188, Giraldus Cambrensis referred to the last British haunt of the beaver (*Castor fiber*) as being on the River Teifi, in Wales. If this is true, and Giraldus was an incorrigible embroiderer, the beaver's extinction might well have been partly due to the removal of riverside woods.

By the 1600s, it is estimated that about an eighth of Ireland was still forested. Within 200 years it had been reduced to a fiftieth, the bulk of the clearances having taken place by AD 1700. The main cause of this decimation was the English plantation settlers and the beginnings of the commercial exploitation of the remaining Irish woodlands. It is not known for certain when the wolf (*Canis lupus*) died out in Ireland but a few may have survived until perhaps 1720 (Fairley 1984). The dramatic loss of woodland cover, with its associated drop in deer populations, on which the wolf preyed, would certainly have speeded up its extinction. In England, the wolf died out about AD 1500, and may have survived for a few centuries longer in the Scottish forests.

Throughout the Middle Ages, an open-field system of farming was typical of the Midlands and central southern England, but elsewhere in the English lowlands the countryside was already characterized by irregular hedge and walled enclosures. As a result of the Parliamentary Enclosures between 1780 and 1820, the remaining open-field systems were replaced by planned, regular fields. The effect of all these enclosures must have been quite dramatic, especially on mammals such as deer, whose ranges became more restricted, and habitats more patchy.

The thirst for good agricultural land encouraged a number of early drainage schemes, and the loss of wetland habitats – especially fens and salt marshes. On the Wash, for example, it has been estimated that some 470 km^2 of salt marsh was reclaimed and enclosed for grazing, mostly before 1800 (See Figure 9.2). When clay-tile drains were introduced in 1800 there was a huge upsurge in drainage activity on wet, clay soils throughout much of England and Wales, and, between 1826 and 1890, over 50,000 km^2 (or 46 per cent) of the agricultural land of England and Wales was under-drained (Robinson 1986). In Ireland, too, there were early schemes to improve bog drainage, although progress was slow. The Bog Commissioners' Act of 1809 had stressed the possibilities of improvements in bog drainage and O'Farrell's Act of 1831 laid stress on water-course drainage by private means. It is estimated that between 1831 and 1922 some 1,520 km^2 of bog were improved (Common 1970).

Very extensive reclamation took place in the Fenland. The first attempt to drain the East Anglian Fenland came with the Romans, who dug open channels to divert the water. But the agricultural use of the Fenland became possible when, in 1629, the fourth Earl of Beford hired the Dutch

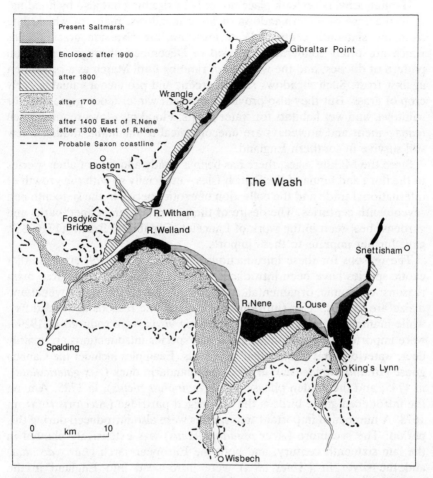

Figure 9.2 Land reclamation on the Wash since 1400. (Modified and redrawn from Cadbury 1987)

engineer Cornelius Vermuyden to begin a wholesale programme of drainage and reclamation. In 1637, when the great drainage scheme began, there were about 3,380 km² of fen. By 1825, some 71 per cent (2,500 km²) had been drained and by the turn of this century only about 3 per cent remained. An unforeseen problem at the time was that of peat shrinkage. Gradually the level of the Fenland has became lower than the drainage channels that were supposed to drain it. In the last 130 years or so, the level of the peat surface on Holme Fen has shrunk by about 4 metres. Drained Fenland provides some of Britain's richest farmland but, when left undrained, fens also provide very species-rich habitats. At Wicken Fen alone some 72 species of molluscs and 212 species of spiders have been recorded.

Habitat conversion took place not only by draining but also by flooding, as in the case of watermeadows or flood meadows, many of which date from the sixteenth century. Watermeadows are riverside grazing-fields which are often deliberately flooded in December, through a grid-iron pattern of ditches, and the water kept running until March as a protection against frost. Such meadows remain green, and produce a valuable early crop of grass. But they also provide important winter feeding-grounds for wildfowl, and wet habitats for water plants. Flood meadows require much management and nowadays are uneconomical to maintain, though a few still survive in southern England.

Since the Middle Ages, there has been a steady addition of alien species to the flora and fauna of the British Isles – especially so with the growth of international trade and the collection of exotic species in the sixteenth and seventeenth centuries. The desire of the gentry for landscaped parks and gardens, best seen in the work of Lancelot 'Capability' Brown (1716–83), gave further impetus to these imports.

The reasons for these introductions varied. Lever (1979) suggests that exotic species have been introduced into the British Isles for three main reasons: economic, ornamental, and sporting. Cereal crops like wheat and maize are alien species (introduced for economic reasons as foodstuffs), while mammals such as the mink (*Mustela vison*) – introduced in 1929 – were imported for their fur. Ornamental species introductions are mainly deer, waterfowl and 'ornamental' pheasants. Examples include: the Canada goose (*Branta canadensis*), in 1660; the mandarin duck (*Aix galericulata*), in 1745; and the golden pheasant (*Chrysolophus pictus*), in 1725. Among the introduced game birds is the red-legged partridge (*Alectoris rufa*), in 1673. A number of important tree species were also introduced during this period. The sycamore (*Acer pseudoplatanus*) was extensively planted in the late sixteenth century, and both the European larch (*Larix decidua*) and the silver fir (*Abies alba*) were introduced into England in the seventeenth century. Exotic tree species were also introduced into Ireland, and the pine was re-introduced about 1700.

The modern period

In the British Isles, some 58 million people live on a land surface area of about 295,000 km², and there are tremendous demands put upon the countryside for all sorts of reasons. Practically all modern farming and forestry practices are harmful to wildlife, and some have led to dramatic declines in the ranges of many species due to habitat loss and disturbance (Nature Conservancy Council 1984). The reason is often very simple: the new agricultural and forest habitats do not contain the basic requirements for many species. For example, the larva of the common blue butterfly (*Polyommatus icarus*) feeds upon bird's-foot trefoil (*Lotus corniculatus*) which disappears when pasture is ploughed, or when treated with a selective herbicide (Nature Conservancy Council 1977).

In her book *Silent Spring*, Rachel Carson vividly portrayed the disastrous ecological effects of over-use of persistent organochlorine insecticides by farmers in the USA, a situation mirrored in many parts of the western world. Organochlorine insecticides such as DDT, aldrin, and dieldrin, are very stable substances and can persist for many years in the environment without breaking down. In the early 1950s they were used extensively as cereal-seed dressing particularly in the arable counties of England. By the mid-1950s ornithologists had noted a marked decline in birds of prey, such as the sparrowhawk (*Accipiter nisus*) and the peregrine falcon (*Falco peregrinus*). The sparrowhawk's range declined spectacularly, and it had all but disappeared from south-eastern England and East Anglia. In 1949 several hundred pairs bred in Norfolk but, by 1965, only a single pair was reported. In 1963 the sparrowhawk was given legal protection under Schedule I of the Protection of Birds Act, 1954–1967.

The collapse of the sparrowhawk's range could not just have been through persecution from gamekeepers, egg collectors, and the like, and another cause was sought. Between 1960 and 1965 a survey showed that predatory birds had much higher levels of organochlorine insecticides in their flesh than other bird species. Sparrowhawks, for example, had up to 3.75 parts per million by weight in breast muscle as compared with less than 0.5 parts per million in moorhens (*Gallinula chloropus*). The differences in levels of insecticide residue were even greater in eggs, and up to 12 parts per million by weight was recorded for sparrowhawks as compared with less than 1.0 for moorhens. Organochlorine compounds are insoluble in water, and are not directly excreted from, but instead are stored in, fatty tissue. In this way they accumulate, and can reach high levels in predators at the top of the food chain. When fat reserves are called upon, as at egg-laying, sub-lethal amounts of these chemicals enter the blood stream, and cause birds to lay eggs with thin, easily broken shells.

All the evidence suggests that loading agricultural ecosystems with persistent wide-spectrum pesticides has harmful effects on the natural flora and fauna. Bryn Green (1981) put the point very evocatively, when he described the effect of herbicides as having 'taken the colour out of the countryside'. The once common, corn marigold (*Chrysanthemum segetum*), cornflower (*Centaurea cyanus*), and corn cockle (*Agrostemma githago*) are now no longer a common sight.

Data on the progressive loss of major habitats in Britain has been collected by the Nature Conservancy Council. There have been catastrophic reductions in some habitats, and unless there is a dramatic change in the way the countryside is managed, the outlook for our flora and fauna is bleak (see Table 9.2).

Table 9.2 Major habitat losses in Britain

Habitat	Recording period	Per cent loss
Lowland heaths	1950–1984	40
Limestone pavements	1950–1976	45 (damaged/destroyed)
Ancient lowland woods	1946–1973	40
Lowland fens	1950–1984	50
Lowland raised mires	1948–1978	60
Upland heaths and bogs	1950–1984	30

Source: Nature Conservancy Council (1984)

The fragmentation of habitats and the reduction in the size of breeding populations are also serious matters which may affect the long-term chances of a species' survival. Because changes in species' abundance are usually not detectable from biogeographical maps, we should bear in mind that a reduction in range and a decline in abundance do not necessarily go hand in hand. In other words, simply because the overall range of species has held up in recent times does not mean that the species is not under some threat.

One method of describing the changes in a species' range – used by Perring (1970) – is to count the number of 10 km^2 grid squares occupied by a species at different dates. For example, if we define a rare flowering plant or fern as one which occurs in fifteen or fewer 10 km grid squares, then, in 1960, some 278 British species qualified, and in aggregate occupied 1,176 squares. However, in 1930, the same species occupied 1,673 squares, indicating a 30 per cent decline in rare species over the period. Perring also estimated the probable causes of decline and extinction in this group of plants since 1800 (see Table 9.3). A thorough review of the definition of species rarity, and other concepts involved in wildlife conservation evaluation, is provided by Usher (1986b).

Table 9.3 The causes for the extinction or decline of rare British flowering plants and ferns, 1800–1960

Per cent	Cause
50	agriculture and drainage
26	natural causes
9	collecting
8	changes in woodland management
7	habitat destruction by building

Source: Perring (1970)

Afforestation

The loss of native timber in the First World War led, in 1919, to the creation in Britain of the Forestry Commission, which was charged with a massive afforestation programme, mainly on the infertile soils of the hills and moorlands of western and northern Britain (Stamp 1969). In Ireland, too, the First World War had an enormous impact, and some 80,000 hectares of woodland were cut, leaving less than half of 1 per cent of the country covered by forest. Planting programmes, both commercial and state, prefer monocultures of fast-growing, mainly alien, species of conifers that do well on poor soils. Pine, Norway spruce (*Picea abies*), and larch were most commonly planted until it was found that sitka spruce (*Picea sitchensis*) grew particularly well in our cool, oceanic climate. Nowadays, this species accounts for about 95 per cent of all forest trees planted in the British Isles but it does not grow well on poor, peaty soils. On such sites the lodgepole pine (*Pinus contorta*) is often planted in its place. In both Eire and the United Kingdom, there have been large afforestation programmes over the last 30 years or so (see Table 9.4). In Eire, the state afforestation programme is the most important land-use change taking place today, and about 8,000 hectares of planting takes place annually. In Britain, the Forestry Commission has a target of about a further 1.5 million hectares of new forests, mainly in the Southern Uplands of Scotland, by the year 2025.

Table 9.4 The extent and ownership of British woodlands, 1970

	Forest area (thousand ha)	Forest area as % land area	Change 1950–1970 (thousand ha)	Ownership (%) State	Private
Eire	275	4	+175	77	23
United Kingdom	1,840	8	+590	40	60

Sources: Peterken 1981; Cabot 1985

Blanket afforestation in the uplands has a serious affect on moorland bird species, and most waders, the skylark (*Alauda arvensis*), and wheatears (*Oenanthe* spp.) do not even tolerate the youngest plantations. In the Southern Uplands the range decline of carrion eaters, such as the raven (*Corvus corax*), is well documented. Vast tracks of sheep-walk have been blanketed with forests, with the result that there has been a decrease in the remains of dead sheep available, and consequently, a decline in the raven also.

Private afforestation schemes in the peatlands of Caithness and Sutherland (better known as the Flow Country) have been particularly well publicized because of the huge tax relief afforded to investors, at least until the budget of 1988. In recent times, four main companies – Fountain Forestry, Economic Forestry, Tilhill, and Scottish Woodland Owners' Association (Commercial) Ltd – have, in total, planted annually about 15,000 ha of the Flow Country with sitka spruce and lodgepole pine. The destruction of these peatland habitats is not only of national importance. The Nature

Conservancy Council estimates that the Flow Country is one of the largest single expanses of blanket-bog habitat in the world (Nature Conservancy Council 1988a).

The afforestation of the Flow Country, a landscape of peaty moorlands and pools, is probably the greatest single threat to birds in the British Isles this century. Nationally important populations of greenshank (*Tringa nebularia*), golden plover (*Pluvialis apricaria*), dunlin (*Calidris alpina*), and merlin (*Falco columbarius*) occur here, as well as scarce species such as the red-throated diver (*Gavia stellata*) and the golden eagle (*Aquila chrysaetos*). The Royal Society for the Protection of Birds estimates that about one-third of the habitat of these species is now owned by forestry interests and about half of this has already been blanketed with conifers. Ironically, the native, pine-beauty moth (*Panolis flammea*), usually resident in Scots-pine woodland, has invaded the lodgepole pine plantations, where the trees are stripped of their needles and die. The forestry companies, not wanting to see their profits diminished have initiated a programme of aerial insecticide spraying, and in 1985 some 5,000 ha were listed for spraying. Presumably, many other insect species are inadvertently killed by such action.

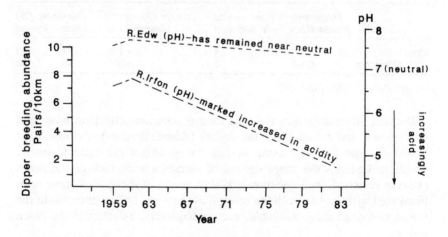

Figure 9.3 Trends in dipper breeding abundance and pH on two Welsh rivers, 1958–83. (Modified and redrawn from Tyler 1987)

One often overlooked effect of afforestation schemes is the way in which they can influence the water chemistry of drainage basins. The further acidification of an already base-poor soil, such as an acid brown earth, is a likely consequence of afforestation with conifers. Ultimately, the effect is felt in the streams draining the catchment. On some Welsh streams, the absence of the dipper (*Cinclus cinclus*), a bird which feeds on aquatic invertebrates such as caddis larvae, is thought to be related to this type of acidification. Figure 9.3 shows dipper populations on two Welsh tributaries of the River Wye. On the River Edw, the pH has remained more or less constant since 1959, but on the River Irfon, the pH has fallen steadily, with the resultant crash in the dipper population. It is no coincidence that the catchment of the Irfon was extensively planted with conifers in the early 1960s.

Changes in woodland management practice

The decline of coppice-woodland management over the last hundred years is thought to be particularly responsible for the range decline of a number of butterfly species. Open areas within the shelter of the woodland are ideal environments for butterflies, but once the coppice is left to grow and a dense shade returns to the forest floor, the habitat becomes unsuitable. One butterfly species whose range has contracted, mostly as a result of the decline of coppice management, is the pearl-bordered fritillary (*Boloria euphrosyne*), whose distribution is shown in Figure 9.4. The decline of this fritillary, which feeds mainly on common dog-violet (*Viola riviniana*), probably started more than 100 years ago but there was a major phase of extinctions in the 1950s and 1960s, exacerbated by the disappearance of the rabbit – whose grazing kept ground flora under control.

Changing agricultural practices

The growth of an agribusiness culture, so typical of much of the British farming scene today, owes much to the Second World War and the necessity then to increase home food production. In Britain the Agricultural Act of 1947 confirmed government policy to encourage agricultural development. As a result, a train of events was set in motion that put many habitats under severe pressure. More recently, the European Community's Common Agricultural Policy with its system of support prices has encouraged grain mountains, butter mountains, huge inputs of fertilizers, and the expansion of farming into marginal areas.

In many upland areas, there is a serious loss of heather-dominated moorland habitats, due to overgrazing by sheep, excessive burning, and, on the moorland fringe, by hill-land improvement. In the United Kingdom, government grants under the Agricultural Improvement Scheme encourage farmers on the moorland fringe to improve its productivity by liming, re-seeding, draining, and ploughing. A similar package of grants for land improvement is available in the Republic of Ireland, under the Farm Development Scheme. O'Sullivan (1983) suggests that increased sheep stocking in the poorer mountainous areas of the republic, encouraged

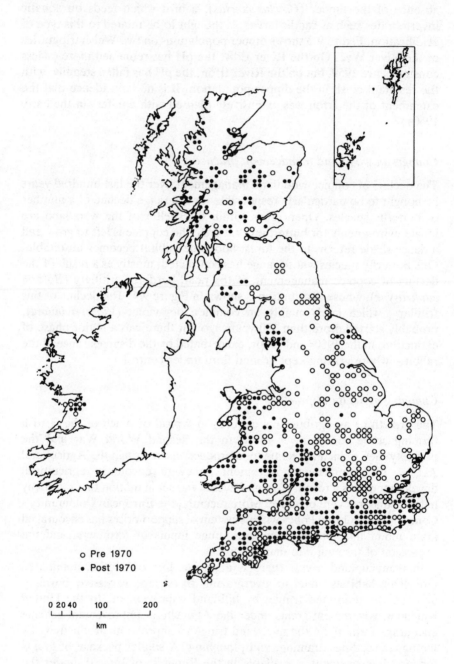

o Pre 1970
• Post 1970

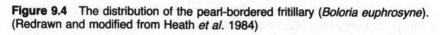

0 20 40   100   200

km

Figure 9.4 The distribution of the pearl-bordered fritillary (*Boloria euphrosyne*). (Redrawn and modified from Heath *et al*. 1984)

by government grants and loans, was, in part, responsible for the decline of the pine marten (*Martes martes*). The pine marten is now mainly confined to woodland and scrub areas of western Ireland, although as recently as 50 years ago it was much more widespread. The reduction in its range is attributed partly to fragmentation of established woodland habitat and also to an extensive poisoning and trapping campaign. To protect sheep and lambs from dogs and other predators, strychnine was put in fresh meat which was then laid on the edges of woodland areas which bordered the mountains. As most of the mountainous regions were common land, the poisoning was often repeated by different farmers. The pine marten takes carrion readily and tends to hunt along the woodland edges more than within woodlands themselves. Hence it was particularly susceptible to the poisoning campaign.

In Britain, the loss of moorland habitats is, in some regions, dramatic indeed (see Table 9.5). In the northern Peak District, Anderson and Yalden (1981) estimated a 36 per cent loss of heather moorland during the period 1913–70 while, at the same time, sheep numbers in hill parishes trebled. Anderson and Yalden suggest that loss of habitat is the prime reason for the decline in red grouse stocks, which are only about a third of their 1930s level. The Cumbrian hills and mountains have suffered even more. There has been something like a 70 per cent loss in upland heather moorland between the 1940s and the 1970s, mostly due to overgrazing, caused mainly by sheep densities being too high, but also by poor shepherding (Nature Conservancy Council 1987). Significant losses of moorland have also occurred both in Northern Ireland and in the Irish Republic.

Table 9.5 Losses of moorland in Britain, 1946–81

| | 1946–81 | | 1976–81 | |
	area lost (ha)	per cent	area lost (ha)	% loss per annum
England	121,957	22	22,352	1.0
Wales	196,793	41	46,316	1.8
Scotland	478,445	11	290,366	1.3

Source: Royal Society for the Protection of Birds (1984)

A number of characteristic moorland birds have suffered widespread population and range declines in Britain, as a result of the reduction in well-managed heather moorland. The merlin, red and black grouse, golden plover, and hen harrier (*Circus cyaneus*) have been particularly affected.

Environmentally Sensitive Areas

Under Article 19 of EC regulation 797/85, a mechanism was provided by which the British Government could designate Environmentally Sensitive

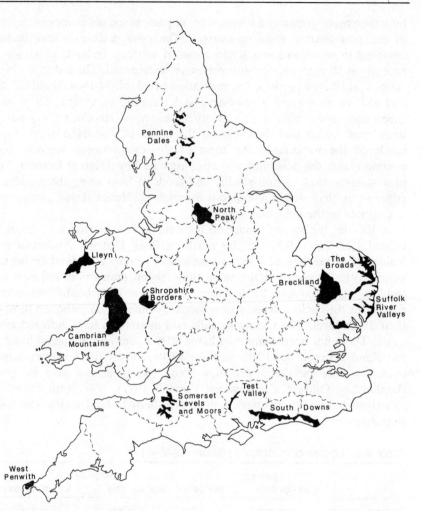

Figure 9.5 The location of Environmentally Sensitive Areas in England and Wales, as of January 1988

Areas (ESAs). These are areas of national wildlife value, which might be damaged or destroyed by changes in existing farming practices. By 1988, twelve ESAs had been designated – including North Peak and the Cambrian Mountains – both of which contain large tracts of heather moorland (see Figure 9.5). Within each ESA, farmers are encouraged to adopt or continue farming practices which achieve conservation as well as agricultural objectives. In return, they receive incentive payments ranging from £30 to £200 per hectare, per year. Participation in an ESA scheme is voluntary and the financial inducements may have to be increased in order to encourage all farmers in designated areas to take part.

In the North Peak ESA, about 23,000 ha of upland moorland come under the scheme. The management plan for participant farmers includes the following:

1 Maintain and improve the quality of the heather by a 12-year burn cycle as laid down by MAFF (Ministry of Agriculture, Fisheries, and Food);
2 Following rotational burning, grazing management should be adjusted so that heather has nil or light grazing and acid grasses have heavy grazing;
3 Summer stocking levels should be such that the heather is not just conserved but extended in both area and quality;
4 The overwintering ewe flock must not be greater than 75 per cent of the 1984–1986 average;
5 Farmers should refrain from improvement work, such as drainage and fertiliser use.

If the ESA schemes are a success, there is some hope of conserving these important upland habitats.

In recent times it is not just the marginal agricultural areas in the uplands that have seen significant habitat losses in the British Isles and in the lowlands both hedgerow removal and land drainage have had profound effects.

Hedgerows

Hedgerows are narrow strips of woody vegetation and their associated organisms; they separate fields and are part of our visual expectation of the countryside. But in addition to their aesthetic qualities they are extremely important as a wildlife habitat. In England, for example, it has been estimated that about 500–600 vascular plant species grow in hedgerows. This habitat is also important for many animals. For example, out of 28 species of British lowland mammals, 21 breed in hedges. Comparable figures for birds are 65 out of 91; and for butterflies 23 out of 54 (Green 1981).

These 'woodland edges without woods' fulfil a variety of habitat roles. Forman and Baudry (1984) suggest that hedgerows act as: a habitat for certain species, particularly edge species; a barrier separating adjacent fields; a source of biotic and environmental influences on adjacent fields; and a corridor for movement of certain species. The species richness of many hedgerows is apparently due to the diversity of microhabitats. The loss of a hedgerow is not simply the loss of one habitat but many. In Table 9.6 at least five distinguishable habitats are occupied by different groups of bird species.

The amount of hedgerow removal is alarming, both in Britain and Ireland. Of the *c.* 804,700 km of hedge in Britain at the end of the Second

Table 9.6 Hedgerow habitats and their use by birds

Habitat	Nesting	Feeding
Upper branches	Carrion crow, rook	Blue tit, chaffinch
Trunks and holes	Barn owl, jackdaw	Treecreeper
Shrubs	Turtle dove, magpie	Fieldfare, robin, redwing
Herbs and brambles	–	Goldfinch, greenfinch
Ground	Skylark	Hedge sparrow, wren, song thrush

Source: Pollard *et al.* (1974)

World War it has been estimated that some 28 per cent (*c.* 225–270 km) had been removed by 1974. In the Republic of Ireland, an average of just under 2 km of hedge per square kilometre may have been lost since 1936 (an average loss of just over 14 per cent) and there is some evidence to suggest that most of this has been since 1973 (Cabot 1985).

According to Hooper (1970b), of the 300 rarest British vascular plants only about ten species are hedgerow plants. However, most of these, such as the balm-leaved figwort (*Scrophularia scorodonia*) and blue gromwell (*Lithospermum purpurocaeruleum*) are not common in the eastern cereal-growing districts of England that have been particularly affected by hedgerow removal. Only fly honeysuckle (*Lonicera xylosteum*), green hound's-tongue (*Cynoglossum germanicum*) and downy woundwort (*Stachys germanica*) are threatened by extinction, although about another fifty species are likely to disappear locally.

In eastern England the loss of hedgerows and traditional hedgerow management is also the main cause for the decline of some butterflies such as the brown hairstreak (*Thecla betulae*) which lays its eggs on *Prunus* leaves – particularly blackthorn. Modern cutters cut back hedges so severely that very high proportions of eggs are removed. This is important in this species, as it is a low egg producer as compared with many other butterflies (Heath *et al.* 1984).

Hedgerow removal, stubble burning, and the application of insecticides have all been blamed for the dramatic decline of the grey partridge (*Perdix perdix*) over the last 50 years or so. Partridges live almost exclusively on arable farmland and nest in the surrounding hedgerows. An average field-size of 10 hectares, surrounded by hedges is their optimal habitat, and provides some 24 km of nesting-cover per 400 hectares of farmland. Even when hedgerows have been maintained, their management by mechanical trimmers discourages partridges from nesting. An annual spring cutting by a trimmer may create neat and tidy hedges, but it also destroys leafy under-cover – so essential to the incubating female.

The corncrake (*Crex crex*), once a common farmland bird, has suffered perhaps more than most bird species as a result of agricultural mechaniza-tion and farming practices. This rather secretive bird nests in fields of grass destined for hay or seed, and only rarely in corn fields. On the farms of today hand scything has been replaced by mechanical mowers, and repeated cutting for silage, rather than the single hay cut, is commonplace. The net effect of these changes, and their disturbance to the corncrake's habitat, has been the progressive decline of this species away from eastern and southern England into strongholds in the north-west of Scotland and Ireland. Recent surveys have shown that even in Ireland the species is declining rapidly, although this might be due to changes in its African wintering and migration areas, rather than the inability of the bird to adapt to different agricultural practices (Cabot 1985).

Land drainage

The reclamation and drainage of lowland fens, bogs, and wet grasslands – all in the name of increased agricultural productivity – is an outstanding factor in the range decline of many species. Soil drainage by mole or tile drains causes the water table to fall, and this can be further encouraged by the widening and deepening of river channels. Once drained soils become aerated, they can be ploughed and dressed with herbicides and fertilizers. The desiccation of the land surface, the physical destruction of wet habitats, and the loading of ecosystems with chemicals is all part and parcel of the production of manicured lowland farmscapes of the British Isles.

In the Somerset Moors and Levels, the drainage of wet meadows and the loss of nationally important bird habitats has reached a critical stage. This patchwork of wet meadows lined by drainage ditches, covers about 65,000 ha. Geologically, the Levels are formed of recent clay deposits which extend inland from the coast only a few kilometres. Behind them come the Moors composed of fen peat, overlying clay (see Figure 9.6). In winter the Levels flood, providing magnificent feeding grounds for wildfowl which gather in their thousands. The dense reed-beds lining the drainage ditches are also the home of nationally rare birds, such as the bittern (*Botaurus stellaris*) which is a legally protected species. These secretive birds, with their booming call, prefer the cover of wetlands with extensive *Phragmites* beds. They once bred in many parts of Britain, and even Ireland, up to about 1840 but, sadly, their distribution is now restricted to about half a dozen areas in England and Wales, as a result of land drainage. The current population is probably only about 80 pairs. The Levels are also one of the few areas in lowland Britain where the otter (*Lutra lutra*) survives.

The 1976–7 RSPB ornithological survey of the Levels recognized five sites of major importance, namely: Kingsmoor, Moorlinch, Tealham/Tadham, West Sedgemoor, and Wetmoor. By 1982, MAFF grant-aided land reclamation had all but destroyed the West Sedgemoor and had severely damaged the rest. The 1983 Water Act requires the English Water Authorities to set up conservation committees and these together

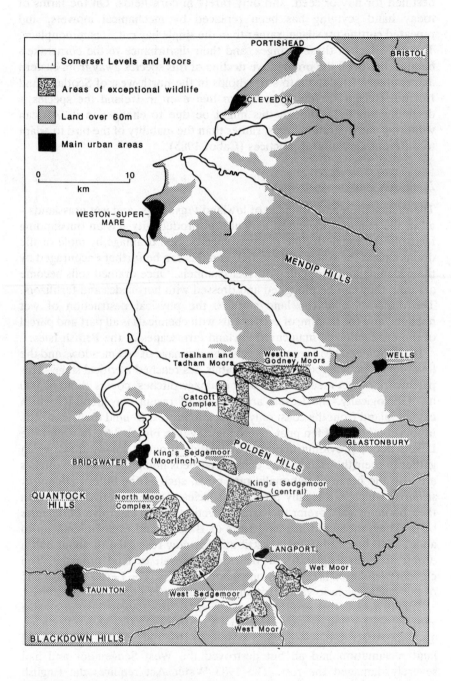

Figure 9.6 The location of the Somerset Levels

with a more enlightened MAFF, should give greater protection to wetland sites.

Although not so intensively farmed, similar land-use pressures also occur in Ireland. On the west coast, for example, drainage threatens the wildlife of a number of turloughs, such as Rahasane Turlough in County Galway. Turloughs are grassy, karstic depressions as much as 1.5 km across and 10 m deep which flood in wet weather. The rapid rise and fall of water gives rise to a very distinctive, zoned vegetation but, above all, is the habitat of the fen violet (*Viola stagnina*) which elsewhere in the British Isles is restricted to the East Anglian fens. Recently, feasibility studies on the drainage of the entire Shannon catchment have been undertaken, and it is likely that, should this scheme go ahead, it will attract huge grants from the European Community.

Of all the wetland habitats in the British Isles, the Norfolk Broads is probably most under threat from the impacts of modern farming and countryside recreation. The Broads, which lie in the middle courses of the rivers Bure, Yare, and Waveney, started life as medieval peat-diggings, which later became flooded. As habitats, the Broads range from shallow navigation channels to broad expanses of open water fringed by extensive reedswamp. But the region also contains large tracts of grazing marshes and productive arable land, whose ecosystems are intimately linked to those of the Broads, as is shown in Figure 9.7.

The Broads are the home of many nationally rare species such as the sharp-leaved pondweed (*Potamogeton acutifolius*) and the dragonfly (*Aeshna isocoles*). Since the early 1950s, and its extinction at Wicken Fen, the Broads have been the last stronghold of the beautiful British race of the swallowtail butterfly (*Papilio machaon britannicus*). This butterfly is restricted to these wetland habitats largely due to presence of its food plant, the milk parsley (*Peucedanum palustre*), which grows in the open fens (Heath *et al.* 1984). There is little doubt that the extinction of the swallowtail at Wicken was due to the drainage of the surrounding farmland and the subsequent decline of its food plant.

The habitats of Broadland are threatened on three broad fronts. First, there are about 10,000 registered boats on the Broads, and inevitably the wash from water traffic disturbs the reedswamps and erodes river banks. Second, reedswamps must be managed to prevent drying out, and the invasion by alder carr. Such management is skilled and labour-intensive, and is now effectively in the hands of amateur volunteers. Third, nutrient enrichment from sewage and farmland run-off has caused severe damage to aquatic life. The eutrophication, which is mainly due to nitrates and phosphates, has given rise to disastrous algal blooms, particularly of *Prymnesium parvum*. In sufficient quantities the toxin released by this alga can wipe out fish stocks and, indeed, this is exactly what happened in Hickling Broad in 1969.

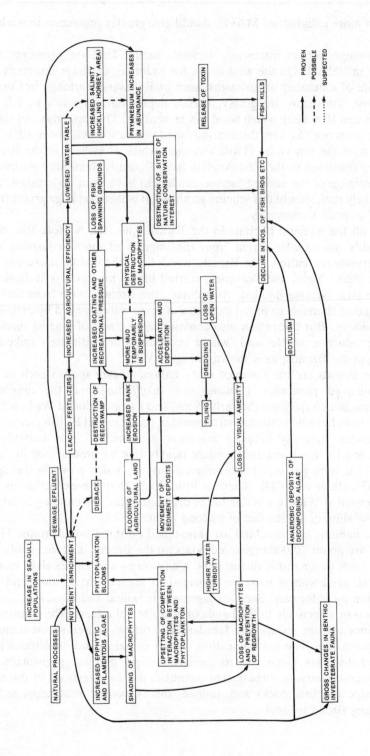

Figure 9.7 Ecological and environmental systems operating in the Broads. (From George 1977)

Habitat protection

The primary objective of conservation is to ensure that our heritage of native flora and fauna remains as large and diverse as possible. To a very large extent, conservation is about habitat protection and we shall conclude this chapter with a brief resumé of the present legal protection afforded to designated wildlife sites. The history of the conservation movement in Britain is given a detailed treatment in *Countryside Conservation* (Green 1981), and can only be touched upon here.

In Britain, early efforts at habitat protection came through the formation of voluntary organizations, such as the Royal Society for the Protection of Birds, the Society for the Promotion of Nature reserves, and the National Trust. The mechanism of habitat conservation afforded by these organizations was one of site ownership, and, prior to the Second World War, there was limited legal protection for either endangered species or habitats. During, and immediately after, the War, a number of committees reported to the British Parliament on the need for nature conservation, and in 1949 the National Parks and Access to the Countryside Act was passed and the Nature Conservancy established. One of the Nature Conservancy's first tasks was the establishment of a nationwide network of Sites of Special Scientific Interest (SSSI). An SSSI is a legal designation, applied to land of special nature-conservation interest, chosen as a result of field surveys, published material, and advice from experts. Notification of the designation of an SSSI is passed on to the land-owner and the local planning authority. The latter body must consult the Nature Conservancy Council if the site is threatened by development (Nature Conservancy Council 1988b). In addition to the establishment of SSSIs, the Nature Conservancy Council was also empowered by the 1949 Act to establish a national series of nature reserves (National Nature Reserves). These were to be the means of protecting, in perpetuity, and through appropriate control and management, the most important areas of natural or semi-natural vegetation with their characteristic flora and fauna (Nature Conservancy Council 1984).

Table 9.7 provides an overview of the various categories of British nature reserves. Many SSSIs are also designated under various UNESCO and EC schemes. For example, there are 31 'Ramsar' wetland sites, 13 UNESCO Biosphere Reserves, and 22 Special Protection Areas in the United Kingdom. Ramsar wetland sites are designated by governments in accordance with the provisions of the 'Wetlands of International Importance' agreement signed at Ramsar, in Iran, in 1971. UNESCO Biosphere Reserves were conceived as part of the Man and the Biosphere programme launched in 1970. Such reserves are protected areas of land or coastal environment representing significant examples of biomes throughout the world, and act as bench-marks for the measurement of long-term changes in the biosphere as a whole (Countryside Commission 1987).

Table 9.7 Types of nature reserve in Britain

Type	Number	Area (ha)
SSSI	4,996*	1,626,796
National Nature Reserve	234	165,478
Nature Conservation Trust Reserve	1,400	44,090
RSPB Reserve	93	43,728
Local Nature Reserve	105	14,371
Wildfowl Refuge	44	11,180
Forest Nature Reserve	11	2,488
Woodland Trust Reserve	102	1,214

Source: Nature Conservancy Council (1988)
* including those in other listed types

In 1981 the Wildlife and Countryside Act considerably strengthened legal provisions for habitat protection in Britain, but not Northern Ireland. The act has four important elements. First, the law protecting the flora and fauna was strengthened with the addition of further endangered species to the lists of protected species. The Act also protects the places occupied by such a species which, in practice, means its habitat. Second, the Act provided for the establishment of Marine Nature Reserves (MNR) which can be designated between high-water mark and the three-mile limit. In 1986 the inshore waters of Lundy Island in the Bristol Channel became Britain's first MNR. A second, on Skomer, off the Pembrokeshire coast, is planned.

Under the 1981 Act, the NCC were given powers to protect any SSSI coming under threat, subject to compensation for the profit foregone in withdrawing proposed agricultural or forestry development. Since 1981 the NCC have completed some 917 management agreements, covering 29,000 ha on SSSIs, and 21,000 ha on National Nature Reserves. Negotiations are proceeding on a further 1,250 agreements. During the period 1985–7, compensation amounting to 11 million was paid out in lump sums and annual payments to site owners and tenants. About 5.5 million has been allocated for such payments in 1988–9. The Act also provided for the protection of limestone pavement, a great deal of which has been quarried as water-worn limestone to adorn garden rockeries and walls.

The 1949 Countryside Act did not apply to the whole of the United Kingdom, and it was not until the Amenities Lands Act (Northern Ireland) of 1965 that legal protection for habitats was secured. The chief power of the 1965 Act was the establishment and management of Nature Reserves, and today there are 44 National Nature Reserves with a total area of about 3,200 ha. The Act also provided for the establishment of some 46 'Areas of Special Scientific Interest' (ASSI) (Forsyth and Buchanan 1982). In 1985 the Amenities Land Act was replaced by the Nature Conservation and Amenities Land Act (N. I.), and the establishment of a Committee for Nature Conservation. This Act brought the status of ASSIs broadly into line with that of the SSSIs in Britain whereby profit compensatory management agreements can be entered into. However, many ASSIs in Northern Ireland are primarily of geological, rather than biological, interest.

In Eire, the Wildlife Act of 1976 provides for the establishment of Statutory Nature Reserves and Areas of Special Scientific Interest (ASSI), both of which are managed by the Forest and Wildlife Service. To date, there are about 1,000 ASSIs, and 44 Statutory Nature Reserves. The total area of the Statutory Nature Reserves is about 8,718 ha. Dublin has the distinction of being the only city to have a Biosphere Reserve within its boundary, at North Bull Island, in Dublin Bay. This internationally important bird habitat was produced as a result of silt accretion due to the completion of a harbour wall in 1825.

Chapter ten

The consequences of urban and industrial growth

In studying the biogeography of the British Isles, it is all too easy to overlook the considerable impacts of urbanization and industrialization, not only in terms of habitat destruction and the consequent displacement of the fauna and flora, but also in terms of the provision of new, 'man-made' habitats. Indeed, some opportunistic species, such as the fox, have taken advantage of the empty niches provided in urban areas, where life among the dustbins is probably as secure nowadays as in the spinney or along the hedgerow.

The biogeographical influences of urban growth and increased industrial activity are often quite far-reaching and are only infrequently contained within the immediate area. The effects of air pollution and acid rain come immediately to mind, but the acquisition of raw materials, energy supplies, water, the disposal of waste products, and the movement of goods and people have all had a measurable impact on the flora and fauna of the British Isles. In ecosystem terms, the inputs and outputs of the urban-industrial system are themselves part of the inputs and outputs of other ecosystems – some adjacent, some far away (Douglas 1981). The complexities of the interactions between urban and rural systems are shown in a schematic way in Figure 10.1. Very few of the pathways referred to in this diagram have been quantified, and one can imagine the difficulties in doing so for a large town such as London, although this has been attempted in other parts of the world. Douglas (1983) provides a detailed examination of such studies in his book *The Urban Environment*.

Recent estimates of land-use change in Britain by the Institute of Terrestrial Ecology (Barr *et al*. 1986) give some indication of the pace of change, particularly in rural areas, but also on the urban fringe, where demands for housing and recreation are important consumers of wildlife habitats (see Table 10.1). More generally, it has been estimated that about 20,000 ha of British farmland are lost annually to urban and industrial spread. To put this last figure into some sort of perspective we should note that it amounts to about half the total area of the nature reserves owned by the Royal Society for the Protection of Birds, a major conservation body.

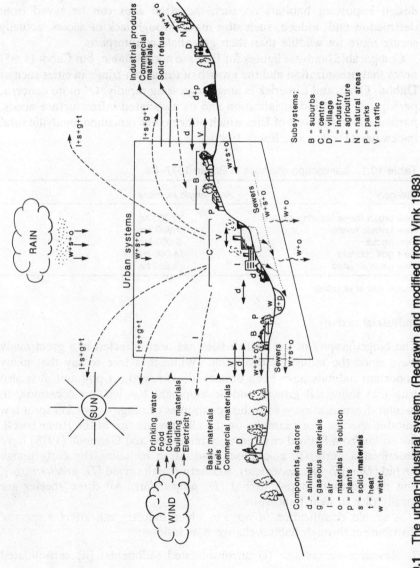

Figure 10.1 The urban-industrial system. (Redrawn and modified from Vink 1983)

Components, factors

d = animals
g = gaseous materials
l = air
o = materials in solution
p = plants
s = solid materials
t = heat
w = water

Subsystems:

B = suburbs
C = centre
D = village
I = industry
L = agriculture
N = natural areas
P = parks
V = traffic

Basic materials
Fuels
Commercial materials

Drinking water
Food
Clothes
Building materials
Electricity

Industrial products
Commercial materials
Solid refuse

Urban systems

RAIN

SUN

WIND

These trends in the rate of urban and industrial growth seem likely to continue, particularly in the south-east of England where the Green Belt land is particularly threatened. But habitats are not only lost to housing. Many industrial developments, such as refineries, power stations, and airports, require vast amounts of land. With sensible planning and good design important habitats on such industrial sites can be saved from destruction and, indeed, such sites may, through lack of access, actually secure more for wildlife than their greenfield counterparts.

Comparable land-use figures for Eire are not available, but Cabot (1985) notes that urbanization and the growth of the urban-fringe in cities such as Dublin, Cork, and Limerick is now progressing rapidly. Of more concern, perhaps, is rural industrialization and its associated infrastructure needs, particularly in the west of Eire which hitherto has remained an idyllic rural backwater (Healey and Ilbery 1985).

Table 10.1 Landscape changes in Britain 1977–84

Category	Area/length increase
New urban-fringe housing estates	39,000 ha
New tarmac roads	4,300 km
New tracks	5,000 km
New golf courses	14,000 ha
New caravan sites	3,800 ha

Source: Barr et al. (1986)

Industrial activity

The biogeography of the British Isles has been affected in a great many ways since the industrial revolution. While it is true to say that many important habitats have been destroyed, damaged, or polluted, it is also true that industrial growth and development has led, on occasions, to habitat diversification – with the possibilities of range expansion for a few suitable species. For example, on the alkali waste tips of the River Irwell, not so far from Salford city centre, Greenwood and Gemmell (1978) have described a herb-rich calcicolous association including the early marsh orchid (*Dactylorhiza incarnata*), southern marsh orchid (*D. praetermissa*), and the northern marsh orchid (*D. purpurella*). All three species are locally rare in west Lancashire.

A simple classification of ways in which industry can affect a species' distribution through habitat change is as follows:

1. Resource extraction: (i) unconsolidated sediments; (ii) consolidated sediments; (iii) water.
2. Pollution: (i) gases; (ii) liquids; (iii) solids.
3. Habitat occupation: (i) building and land development.
4. Importation: (i) raw materials and aliens.

Resource extraction

The mining of unconsolidated deposits – such as clays, sands, gravels, and peat – is a mixed blessing from a biogeographical point of view. Generally speaking, the actual physical consumption of land produces hollows which flood when near the water table. If the extraction takes place in agricultural floodplains there may be a considerable increase in the variety of habitats available, and a concomitant increase in the number of species (see Table 10.2). Man-made habitats have been very beneficial to the range extension of some species, such as the little ringed plover (*Charadrius dubius*). On the Continent, this bird nests on gravel banks in river valleys, and was unknown as a breeding species in Britain before 1938. Since then it has spread widely throughout much of eastern England where it is normally associated with the margins of disused gravel pits.

Table 10.2. A comparison of the number of habitats and number of bird species on the Ouse floodplain around Milton Keynes

	Great Linford gravel pits	'Normal floodplain'
Habitat types	14	5
Bird species	123	89

Source: Kelsey (1975)

In Eire, peat is an important fuel both for domestic and industrial purposes. In 1946, commercial exploitation was nationalized and the industry run by Bord na Mona (Irish Peat Board), which owns some 81,000 hectares of bog. Most mechanical peat harvesting has concentrated on the raised bogs of the central lowlands, but, recently, blanket peats in north-west Mayo have been exploited. Production figures are quite staggering. In 1964–5, Bord na Mona produced 336,798 m^3 of peat, and by 1981–2, this had risen to 1,191,241 m^3. Forecasts suggest that by the year 2020 most Irish-raised bogs and many blanket bogs will have disappeared through exploitation, leaving nothing but a thin layer of residual peat, and an enormous problem of landscape restoration. Such massive habitat destruction is almost unprecedented in the British Isles and at least ten bog areas of international importance totalling 12,767 hectares are under threat. Runoff from the disturbed bog surfaces takes with it vast quantities of suspended peat solids, which accumulate in lakes, where it interferes with fish spawning beds (Cabot 1985).

Raised bogs, rare elsewhere in Europe, are important habitats for a number of moorland birds, such as the golden plover and snipe, but above all for the Greenland race of the white-fronted goose (*Anser albifrons*). Up to three-quarters of the world's population overwinters in Eire, and the loss of raised bog habitats is the main cause of their population decline in recent years.

Another serious habitat threat is the quarrying of limestone, particularly on the Magnesian and Carboniferous limestone outcrops (Ratcliffe 1974). The Magnesian Limestone escarpment stretches from Nottingham northward to Durham, and is nowhere very high. From a floristic point of view, interest lies in fact that calcareous southern elements – such as the bee orchid (*Ophyrs apifera*) and the upright brome (*Zerna erecta*) – reach further north on these dolomitic limestones, than on the Carboniferous Limestone outcrops to the west.

Although the habitats of some species, such as the pasque flower (*Anemone pulsatilla*), have been totally destroyed by quarrying, many disused quarries on the Magnesian Limestone outcrop act as refuges for plants and animals of the surrounding semi-natural, calcareous grasslands. Indeed, such is the botanical interest of some of these quarries, such as Raisby Quarry and Fulwell Quarry, near Durham, that they have been designated SSSIs (Davis 1981).

Carboniferous limestone outcrops are quarried extensively in the Pennines, in north and south Wales, and the Mendip Hills. In these areas a number of important woodlands have been largely or totally destroyed as a result, and others threatened. For example, quarrying permission dating back to 1946 covers some 140 ha of Asham Wood in the Mendips. This is floristically the richest of the Mendip ashwoods with a number of local plant species such as Solomon's seal (*Polygonatum multiflorum*), small teasel (*Dipsacus pilosus*), and meadow saffron (*Colchicum autumnale*). The wood is also the habitat for the wood white butterfly (*Leptidae sinapsis*) and a rare snail, the mountain bulin (*Ena montana*).

Many of the recently glaciated outcrops of Carboniferous limestone bear limestone pavements, whose runnelled surfaces are separated into massive clints by the weathering of vertical joints, locally known as grikes. The humid microhabitat of the grikes is the home of many species of fern and flowering plant. In Britain, the vary rare angular Solomon's seal (*Polygonatum odoratum*), and the rigid buckler-fern (*Dryopteris villarii*) are virtually restricted to such sites. The removal of the runnelled surfaces of pavements for lime, crushed aggregate, and as decorative rockery stone, has been quite extensive. A comprehensive survey of limestone pavements in Britain by the Nature Conservancy Council in the mid-1970s estimated that, of the 2,150 ha of pavement, 97 per cent showed signs of damage and only 13 per cent were more than 95 per cent intact. Even in the floristically-rich karst of the Burren on the west coast of Eire there is evidence of widespread damage to limestone pavements (Goldie 1986).

Water is a major resource requirement for both industries and cities and the provision of adequate supplies has had both far-reaching and damaging consequences for many habitats, and their associated fauna and flora. Terrestrial habitats are simply drowned by the creation of reservoirs whose frequently fluctuating water levels create very precarious marginal

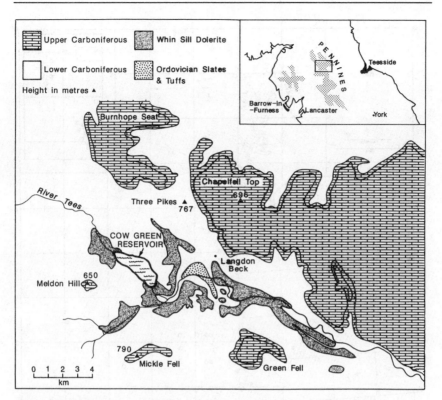

Figure 10.2 Location of the Cow Green Reservoir in upper Teesdale

habitats. Ratcliffe (1984) notes the loss in this manner of the rare slender rush (*Juncus filiformis*) from the rocky shores of Thirlmere in the Lake District.

The best publicized case of an important habitat being partially drowned by the creation of a water-body was that of the Cow Green Reservoir in upper Teesdale. This reservoir was proposed by the Tees Valley and Cleveland Water Board, and the Northumbrian River Authority, to supply the extra water – needed mainly for the ICI chemical works – on Teesside. The chosen reservoir location had long been recognized as one of great importance botanically, for here there are extensive areas of limestone grassland on well-drained calcareous soils, developed over granular, sugar limestone. There are also well-developed, wet, calcareous flushes and acid bogs (Clapham 1978). After a good deal of hassle, permission was given for a reservoir, and in 1970 a dam was built across the River Tees above Cauldron Snout (see Figure 10.2). As a result, a significant proportion of the calcareous habitats of rare, montane species such as Teesdale violet (*Viola rupestris*) and kobresia (*Kobresia simpliciuscula*) was flooded, drowning the species. Important bog communities were similarly affected,

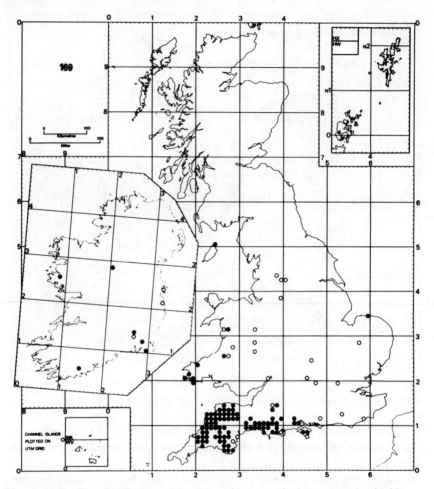

Figure 10.3 The distribution of the lichen *Usnea articulata*, a species highly sensitive to air pollution. Open circles = pre-1960. (Redrawn and modified from Seaward and Hitch 1982)

and there was also a loss of habitats containing the very local, tall bog-sedge (*Carex paupercula*), and the rare, northern water-sedge (*Carex aquatilis*). As is often the case, the economic forecasts were wildly wrong, and the industrial growth on Teesside has not materialized.

Pollution

The release of man-made toxic wastes into the environment is a particularly pernicious consequence of the growth of industry and population over the last hundred years or so. The effects of such pollution can be both widespread, as in the case of air pollution, or very local, as in the case of a toxic waste tip. In either case, species which are sensitive to the levels and persistence of the toxicity are displaced from the polluted area. Of course not everywhere in the British Isles has suffered to the same extent from the effects of pollution, and, broadly speaking, both Eire and Northern Ireland, with their rural economies, are much less affected than Britain. The most widespread form of air pollution is caused by sulphur-dioxide emission (see Table 10.3), which has brought about major changes in the distributions of some taxa, especially lichens. All lichens are a symbiotic union of an alga and a fungus. The fungal partner, or mycobiont, usually forms the bulk of the lichen thallus. The algal partner, or phycobiont, is usually a green or blue-green alga, which in many cases, can also be free-living. Lichens are a particularly successful group of organism and occur naturally in an extremely wide range of habitats and substrates with about 1,380 species known in the British Isles.

Table 10.3 Some comparative European sulphur dioxide emission levels

Country	1980 SO₂ emissions (10³ tonnes)	Area (10³ km²)	Annual SO₂ emission	
			Tonnes/km²	Tonnes/10³ population
Portugal	170	76	2.2	17.2
Ireland	217	70	3.1	63.8
France	3,270	544	6.0	60.8
United Kingdom	4,680	244	19.1	83.5
Czechoslovakia	3,100	128	24.2	206.7
German Democratic Republic	4,000	108	37.0	250.0

Source: Cabot (1985)

Most species of lichen are particularly sensitive to increased levels of sulphur dioxide, which causes the breakdown of chlorophyll *a* to phaeophytin *a* in the algal cells. Sulphur dioxide is released into the atmosphere when fossil fuels are burned, and in the smelting of metal ores; some is also released naturally into the atmosphere from volcanoes. The natural background atmospheric concentration of this gas is between 0.28 and 2.8 µg m⁻³, but in the 1952 London smog levels went as high as 35 100 µg m⁻³ (Hawksworth and Rose 1976).

Over much of Britain, the range of many lichen species is contracting away from the industrial and urbanized zones, where the atmospheric concentration of sulphur dioxide is high. Not all species are equally sensitive to this type of pollution, but after taking into account other variables, such as substrate, it has been discovered that there is a good spatial association between particular species and the concentration of sulphur dioxide in the atmosphere. In one scheme, seven lichen zones are broadly related to mean winter sulphur-dioxide levels (see Table 10.4).

Table 10.4 The Advisory Centre for Education's (ACE) pollution zone scale

Zone	Lichens (and some mosses)	Approximate mean winter SO_2 levels ($\mu g \, m^{-3}$)
0	Lichens absent on trees but the green alga *Desmococcus viridis* may be present	+ 170
1	*Lecanora conizaeoides* on trees and stone	150–160
2	*Xanthoria parietina* appears on concrete, asbestos, and limestone	125
3	*Parmelia saxatilis* and *P. sulcata* appears on acid stone, the moss *Grimmia pulvinata* occurs on limestone and mortar	100
4	Leafy (foliose) species, e.g. *Hypogymnia physodes* on trees	70
5	Shrubby (fruticose) species, e.g. *Evernia prunastri* on trees	40–60
6	Species of the genus *Usnea* became abundant	35

Source: Modified from Gilbert (1974)

The distribution of corticolous lichens (those occurring on trees) has been extensively studied, and, as early as 1859, L. H. Grindon suggested that their decline in south Lancashire was due to the increase in air pollution in the area. In England, it is now thought that about ninety species have become rare or extinct where mean winter sulphur dioxide levels are over *c.* 65 $\mu g \, m^{-3}$. A good example of this group is *Usnea articulata* which, although once quite widespread, is now mainly confined to the south-west of the British Isles (see Figure 10.3), where mean winter sulphur-dioxide levels are less than about 30 $\mu g \, m^{-3}$. In contrast, species such as *Lecanora conizaeoides* and *Parmeliopsis ambigua* have increased their range dramatically as a result of air pollution. Evidently, *Lecanora* has spread rapidly over much of Britain since about 1860 and is common where winter sulphur-dioxide levels are between 50 and 150 $\mu g \, m^{-3}$.

Local studies illustrating the effects of smoke and sulphur dioxide pollution levels on the zonation of lichens, have been carried out in the lower Tyne valley by Gilbert (1970), and in Belfast by Fenton (1964). Goodman and Roberts (1971) describe essentially similar air pollution effects on the distribution of the epiphytic moss *Hypnum cupressiforme* around Swansea where, at the time of sampling, non-ferrous smelters also delivered large quantities of copper, zinc, and lead into the air.

The pollution of water refers to the addition of any substance, including heat, which alters its chemical or physical properties. These effects are not, of course, limited to freshwater systems, and there is extensive pollution of estuarine and coastal waters around the British Isles. Fertilizer and slurry runoff from farms, untreated domestic sewage, and a veritable cocktail of industrial waste products enter British river courses every day, and eventually find their way to the open sea. Many marine habitats are also becoming polluted from the huge quantities of

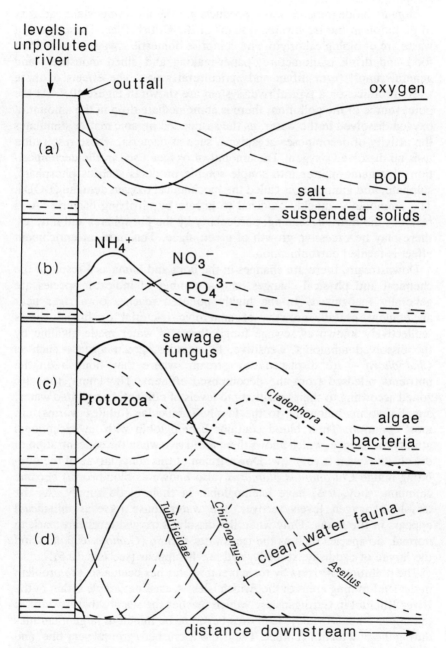

Figure 10.4 Effects of organic biodegradable effluent on (a) physical changes; (b) chemical changes; (c) flora; (d) fauna of a river. (Redrawn from Hynes 1958)

waste emptied from bilge tanks, by ships plying some of the busiest shipping lanes in the world.

Organic biodegradable waste products are the most important category of pollution in the freshwater systems of the British Isles. These types of waste are of biological origin, and comprise domestic sewage, wastes from food and drink manufacture, paper-making and allied industries, and general runoff from urban and agricultural areas. The effects of these waste products on a typical river system are shown in Figure 10.4. At the point source of the pollution, there is an immediate drop in the amount of oxygen dissolved in the water, as the suspended organic matter stimulates the activity of decomposer organisms, such as bacteria, whose respiration uses up dissolved oxygen. The amount of oxygen used in the decomposition of organic matter into simple soluble products such as phosphate, sulphate, and ammonia, is called the biochemical oxygen demand (BOD). The ammonia is quickly oxidized to nitrate by nitrifying bacteria. As a result of the nutrient loading, particularly by the phosphates and nitrates, there may be excessive growth of green algae. This nutrient enrichment effect is termed eutrophication.

Downstream, there are changes in the flora and fauna as a result of the chemical and physical changes, and a number of indicator species are generally recognized. In the highly polluted reaches of a river near an effluent outfall, growths of anaerobic bacterial or fungal slimes (collectively known as sewage fungi) thrive in water made alkaline by the dissolved ammonia. Sensitive, filamentous, green algae – such as *Cladophora* – are displaced downstream, where they flourish on the nutrients released from the decomposed effluent. The fauna, too, are zoned according to their sensitivity to levels of effluent. In polluted water, annelid worms belonging to the family Tubificidae (tubifex worms) are most common. Their blood contains haemoglobin with an exceptional affinity for oxygen, which allows them to survive when the concentration of dissolved oxygen is very low. Downstream of this zone, larvae of the non-biting midge *Chironomus plumosus* (also known as bloodworm) become common; they, too, have haemoglobin in their blood supply. As the dissolved oxygen levels recover, the water louse (*Asellus aquaticus*) appears in the fauna. Only when the dissolved oxygen levels are back to normal, do species such as the freshwater shrimp (*Gammarus pulex*) and the larvae of caddisfly, stonefly, and mayfly appear (see Fig. 10.5).

The pollution of rivers by toxic metal wastes has been a severe problem in the lead-mining areas of the British Isles. A classic example is that of the River Rheidol in Cardiganshire, which reaches the sea at Aberystwyth. In the upper Rheidol valley, lead ores began to be mined in large quantities during the nineteenth century. The ore was crushed, ground very fine, and then washed with water to extract the lead. The washings, with their lead-laden suspended load were collected in settling pits, and then dumped on

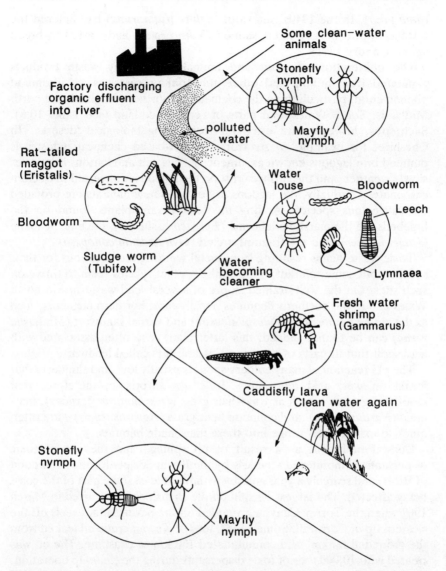

Figure 10.5 Down-stream changes in the fauna of a polluted river. (Redrawn and modified from Simkins and Williams 1981)

the land from where the lead leached into the river. In 1919, when the mining industry was petering out, the Rheidol was quite devoid of fish, invertebrates, or plants, apart from lead-resistant filamentous green algae. Mining finally ended in 1927, and there was a rapid recolonization of the river system. By 1925, the first sticklebacks (*Gasterosteus aculeatus*) had been found, and, by 1932, people were fishing for brown trout (*Salmo trutta fano*). In the 1940s, sea trout (*Salmo trutta trutta*) had entered the system, and, a decade later, salmon (*Salmo salar*) had started to breed again (Condry 1981).

The obliteration of plant and animal habitats by waste products generated from mining and other industrial processes is a widespread phenomenon, particularly in the coalfields of northern England, the north Midlands, South Wales, and parts of central Scotland (see Table 10.5). Such sites, however, are often far from being biological 'deserts'. In Cheshire, for example, waste from the salt-based chemical industry is pumped into lagoons known as 'limebeds'. The salt and various impurities slowly weather, and the surface crust on the lagoon is converted to calcium carbonate. Gradually, the lagoons dry out and new habitats are provided for calcicolous species. Among the locally rare plants found on the limebeds are the marsh helleborine (*Epipactis palustris*), the blue fleabane (*Erigeron acer*), and the fragrant orchid (*Gymnadenia conopsea*).

Toxic waste heaps resulting from metal working are notorious for their sterility, but some tolerant, opportunistic, plant species seem to thrive on such sites. On the well-drained slopes of several lead-waste tips in north Wales and in the northern Pennines, locally rare, northern elements, such as the alpine pennycress (*Thlaspi alpestre*) and vernal sandwort (*Minuartia verna*) can be found. Indeed, this latter plant is so often associated with lead spoil that in parts of northern England it is called leadwort.

The pH reaction of many colliery spoils is usually low, and similar to that found on wet, acid moorlands. It is not surprising, therefore, that moorland grasses – such as wavy-hair grass (*Deschampsia flexuosa*), mat-grass (*Nardus stricta*), and common bent-grass (*Agrostis tenuis*) – are often quick to extend their range into these man-made habitats.

Coastal pollution, as a result of oil spillage and the use of toxic dispersants, is monitored regularly by the Royal Society for the Protection of Birds, and scarcely a year goes by without at least some part of the coast being affected. The largest oil spill on the British coast occurred in March 1967 when the Torrey Canyon grounded on the Seven Stones reef, off the western tip of Cornwall. Some 14,000 tons of Kuwait crude oil leaked from the grounded tanker, and contaminated 150 km of coastline. The oil was treated with 10,000 tons of toxic dispersants during the clean-up operation. As it turned out, the dispersant was extremely toxic to both plants and animals living in the shore zone. Practically all animal life close to dispersant spraying was killed, as were most algae (Southward and Southward 1978). In less than a year rocks sprayed with dispersant were covered by the green alga *Enteromorpha*, and, soon after, the brown algae *Fucus vesiculosus* and *Fucus serratus* became re-established. A settlement of limpets and other grazing animals followed, leading to the eventual reduction and loss of the brown algae. Within 10 years, most polluted sites had recovered their species richness. One species, the rare, warm-water,

Table 10.5 Major types of waste and their location in Britain

Type of waste	Stockpile (M tonnes)	Area (ha)	Production (M tonnes/year)	Principal locations
Colliery spoil	3,000	15,000	50	Coalfields of N. and NE. England, Midlands S. Wales, C. Scotland and Kent
Blast furnace slag	?	?	9	S. Wales, Corby, Sheffield, Scunthorpe
Steel making slag	?	?	4	(as for blast furnace slag)
Metal mine spoils	?	?	0.46	Cornwall, N. and C. Wales, Derbyshire
Metal smelter wastes	?	?	0.24	Swansea (Zn, Cu), Widnes and Walsall (Cu)
Power station ash	?	?	10	Nationwide
Slate waste	300	?	1.2	Wales, Lake District, Cornwall, Scotland
China clay waste	280	800	22	South-west England
Chemical wastes	?	?	?	Merseyside and Cheshire (alkali and salt wastes) S. Wales and Fife (red mud from alumina manufacture)

Source: Modified from Gemmell (1977)

hermit crab (*Clibanarius erythropus*) failed to re-establish its former range in the area. Evidently, the British population of this crab species has a rather precarious status and is maintained by occasional planktonic larvae derived from warmer waters on the Brittany coast.

Building and land development

The consumption of land by and for various types of industrial and commercial development is particularly severe on estuarine sites in the British Isles. With the approach of winter in northern Canada and northern Europe, vast numbers of waders migrate south to the temperate estuaries of the British Isles. Many of these muddy habitats are of outstanding international importance for wading birds and also provide crucial stop-over sites for birds migrating between their northern breeding sites and overwintering areas in southern Europe and west Africa. From the palaearctic breeding areas come the grey plover (*Pluvialis squaterola*), bar-tailed godwit (*Limosa lapponica*), and dunlin (*Calidris alpina*); from the nearctic biogeographial region come knot (*Calidris canutus*), sanderling (*Calidris alba*), and turnstone (*Arenaria interpres*).

Estuaries provide excellent opportunities for industrial and commercial expansion. In recent years, estuarine land has been claimed for shipping terminals and the storage and refining of oils at Teesmouth, on the Thames estuary, on the Solent, at Milford Haven, and at Nigg Bay on the Cromarty Firth. In fact, a gamut of industrial and commercial pressures are threatening British estuaries – ranging from possible airport development on the Maplin Sands, which is a major European overwintering station for the brent goose (*Banta bernicla*), to the development of an alumina smelter and major coal-fired power station on the Shannon estuary, by far the most important site for overwintering waders and wildfowl in Ireland. The Royal Society for the Protection of Birds has investigated the impacts of industrial and urban pressures on important bird habitats. Figure 10.6 illustrates the variety of habitat pressure due in large part to conflicting land-use demands, particularly along the coast.

In percentage terms, the scale of habitat loss in some estuaries is enormous. At Teesmouth, for example, intertidal mudflats such as Seal Sands, Bran Sands, and Coatham Sands, have been reduced from about 2,500 ha in the 1930s, to a mere 149 ha in 1976 (Cadbury 1987). Much more dramatic losses of estuarine habitat are just on the horizon, however, as the threat of barrages for tidal-power generation becomes a reality. Major habitat changes are likely as a result of barrage construction. Inside a barrage, tidal amplitude is reduced and the exposure of inter-tidal mudflats restricted; pollutants will probably accumulate, and there is increased disturbance through recreational activity. In May 1988 government planning permission was given for a £450-million barrage across the

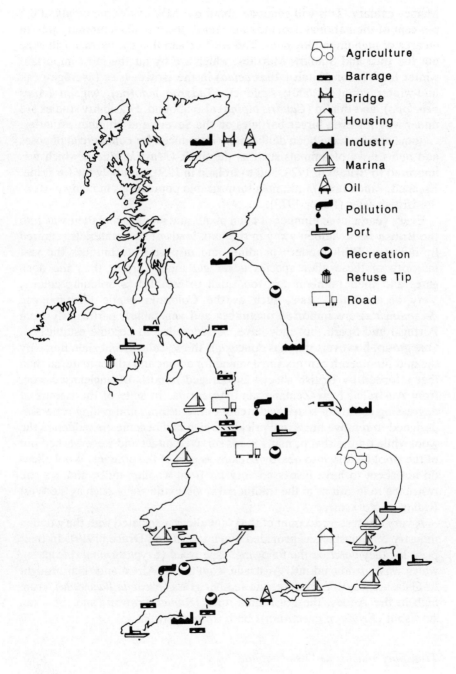

Figure 10.6 Major environmental impacts on coastal and estuarine sites in the United Kingdom. (Redrawn and modified from Cadbury 1987)

Mersey estuary. This will generate about 621 MW of electricity, about 0.5 per cent of the national demand, and create about 5,000 temporary jobs in an area of high unemployment. The RSPB claim that the barrage will wipe out the Ince and Stanlow Marshes, which are by far the most important winter habitat for pintail (*Anas acuta*) in the British Isles (average peak mid-winter count 11,600); shelduck (*Tadorna tadorna*), wigeon (*Anas penelope*), and dunlin (*Calidris alpina*) also abound. Feasibility studies are under way for even larger barrages on the Severn and Humber estuaries.

Some species have been deliberately introduced for commercial puposes and one thinks of animals such as the mink (*Mustela vision*) which was imported to Britain in 1929, and to Ireland in 1950, to supply the fur trade. Escaped, ranch-bred mink, now form sizable populations in many parts of the British Isles (Lever 1979).

Every year a large number of alien plants and animals find their way into the British Isles, hidden away in the bulk loads of raw materials required by industry. With modern handling and inspection techniques the vast majority of these alien species never get much further than the dock gates and their numbers are too small to establish a breeding colony. Very occasionally, pests, such as the Colorado beetle (*Leptinotarsa decemlineata*) are imported on timber and vegetables, particularly from Portugal and Spain, but they have, so far, failed to become established. One group, however, that has done so, is the so-called wool alien flora. By the mid-nineteenth century the demand for a finer and silkier material than that afforded by British sheep, encouraged imports of uncleansed wool from Australia, New Zealand, and Argentina. In spite of the rigours of degreasing in vats of soap, acid treatment, crushing, and rolling processes designed to remove burrs and prickles which had become embedded in the wool while on the sheep, many seeds survived intact and were washed out of the wool factory into nearby streams as part of the effluent. Wool aliens do not seem to have dispersed very far from woollen mills, and are still restricted to locations in the traditional wool textile areas, such as the West Riding of Yorkshire.

A very interesting account of the plant aliens associated with the woollen industry of Galashiels is provided by Hayward and Druce (1919). In their plant list they describe the following: cape weed (*Cryptostemma calendula*), which was introduced into Australia from South Africa and then brought to Galashiels and Yorkshire; the feather grass (*Nassella flaccidula*), from high in the Andes; the Australian dock (*Rumex brownii*) and the swan storksbill (*Erodium cygnorum*), both from Australia.

Transport systems as linear habitats

The development of trackways, roads, canals, and railways has had a considerable influence on the British flora and fauna. These communication systems form a vast network of habitat corridors, and provide almost unrivalled opportunities for rapid colonization and range extension (Lousley 1970). Of course, the construction of this network was not without a cost, as a great deal of land was necessary. It has been estimated, for example, that 1 km of motorway occupies about 5 ha of land. On the other hand, many new habitats were created – often stretching for hundreds of kilometres – connecting settlement with settlement, and the town with the countryside. The present extent of the British transportation systems is shown in Table 10.6

Table 10.6 The length of the road, rail, and canal systems in the British Isles (km).

	Road	Rail	Canal
Eire	92,303	1,876	188
Northern Ireland	23,731	357	–
Britain	348,344	17,068	3,200
Total	464,378	19,301	3,388

Source: Europa Year Book (1988)

Roads

A close inspection of a transport system reveals more habitat diversity than one might expect (Figure 10.7). Such linear habitats may not seem imposing, until it is remembered that they are often highly connected, often free from direct human interference, and are frequently refuges for plants and animals which were once common in the adjacent farmland. Roadside verges in lowland, arable England, for example, contain some of the last vestiges of the grasslands which existed before the modernization of agriculture (Perring 1969). They also contain at least 27 of Britain's 300 rarest plant species, and in some cases the roadside is now their main habitat (see Table 10.7). In the case of hare's ear (*Bupleurum falcatum*), a roadside verge in Essex was its only location in Britain, until that was destroyed by road works in 1955. A similar fate fell upon the chalkhill blue (*Lysandra coridon*), a local butterfly, characteristic of old calcareous grasslands in southern England. For many years, its range extended north to a road verge near Ancaster in Lincolnshire, but this colony was destroyed in 1970. These two examples highlight the critical role of management on the stability of roadside verge habitats. Nowadays, many highways authorities have agreements with their county naturalist trust to limit herbicide treatment on important verges.

The range extension of maritime plants along roadside verges in north-east England has been investigated by Scott and Davison (1982). The application of de-icing salt to major roads in this region has created highly

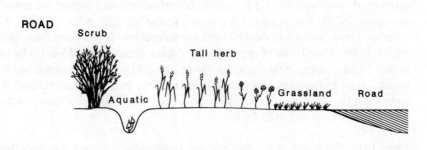

ROAD

Scrub

Tall herb

Aquatic

Grassland

Road

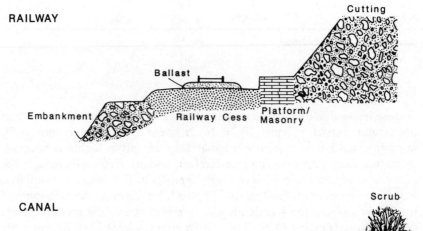

RAILWAY

Cutting

Ballast

Embankment

Railway Cess

Platform/
Masonry

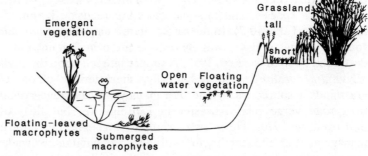

CANAL

Scrub

Grassland

tall

short

Emergent
vegetation

Open   Floating
water vegetation

Floating-leaved
macrophytes

Submerged
macrophytes

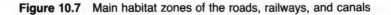

Figure 10.7   Main habitat zones of the roads, railways, and canals

Table 10.7 Rare plants mainly occuring on roadside verges

Beta trigyna (beet)
Linum anglicum (perennial flax)
Melanpyrum arvense (field cow-wheat)
Muscari atlanticum (grape hyacinth)
Orobanche caryophyllacea (clove-scented broomrape)
Phyteuma spicatum (spiked rampion)
Pyrus cordata (Plymouth pear)

Source: Modified from Perring (1969)

saline bare ground adjacent to the road. Maritime grasses, such as the reflexed poa (*Puccinellia distans*), are now widespread on these saline habitats as are other saltmarsh species, such as sea aster (*Aster tripolium*), the shore orache (*Atriplex littoralis*), and common scurvy-grass (*Cochlearia officinalis*). Scott and Davison (1982) have recorded the roadside distribution of several of these species and their distribution map showing the inland range extension of *Puccinellia distans* is shown in Figure 10.8, together with its general distribution as recorded by Perring and Walters (1962).

Evidently, the range extension of this maritime flora is quite recent. The heavy application of de-icing salt on trunk roads only began in the mid-1960s, and much of this flora in north-eastern England is restricted to those roads built after 1967. At several sites, such as Holy Island in Northumberland and Cowpen Marsh south of Hartlepool, there are roads which cross saltmarshes, and these eventually link up with the A1, M1, and M62 trunk roads. The seeds of the saltmarsh plants might have dispersed in the slip-streams of vehicles or perhaps became temporarily embedded in tyre treads. Several roads in eastern Kent show a similar pattern of invasion. Vehicle slip-streams also probably helped the rare, sheet-web spider (*Tegenaria agrestis*) to travel from its only previously recorded site in the New Forest to the concrete columns of the Gravelly Hill Interchange, on the M6 near Birmingham.

Railways

The growth of the railway network in the British Isles was quite spectacular. The first public railway bill was presented to the British Parliament in 1801, and by 1825 there were 669 km of track; 100 years later, in 1925, this had increased to 84,107 km. The immediate effect of the construction of railways was the destruction of existing habitats, and the creation of new ones (see Figure 10.7). Railway habitats can be divided into cess and verge. The cess is the freely draining cinder bed, over which the track bed ballast (the permanent way) and rails are laid. Alongside the cess are various masonry habitats, such as bridges, concrete posts, platforms, and buildings. Verges are either cuttings which slope up from the cess, or embankments which slope down to the surrounding land. The youthfulness of both cuttings and embankments usually means that the soil profiles are poorly developed, although the addition of topsoil to embankments may give the opposite impression. The total area of verge habitats for rural lines in the British Rail network is approximately 30,678 ha.

A number of plant species occur remarkably often in the freely draining habitats of the cess and verges, and have been called 'railway species'; together they constitute the *railway flora*. Early studies of the railway flora were those of Dony (1955) in Bedfordshire, and Messenger (1968), in the now defunct county of Rutland. Messenger concluded that the railways of Rutland (now part of Leicestershire) were among the richest botanical areas in that county. He and his co-workers found 372 plant species on railway property, and only 20 of the hundred 4 x 4 km grid squares surveyed contained a greater number of species.

The most thorough survey of the railway flora is that recently completed by the Institute of Terrestrial Ecology, for the whole of the British Rail network (Sargent 1984). This study recognized some nineteen species as occurring in more than 5 per cent of the 893 stretches of track examined (see Table 10.8). The most common plant recorded, and *the* railway species *par excellence*, was the rose-bay willow-herb (*Chamaenerion angustifolium*). Until 100 years ago, this was a relatively rare plant found on rocky scree slopes and in woods. Its change in status, from a rarity to a very abundant plant, was probably due to the run-down in rail maintenance during the inter-war period when there were many open derelict habitats in the little used goods yards due to increased road transport, and the depressed state of the economy (Gulliver 1980).

Regular herbicide treatment of the cess keeps it clear of most species, although several winter annuals – such as spring whitlow grass and thale cress – are quite common. These species are able to complete their life-cycle before chemical spraying takes place in the early summer, and lie dormant in the seed phase over the summer months.

The role of railways in the spread of the Oxford ragwort (*Senecio squalidus*) is well known. This species is a native of Sicily and southern Italy, and has been cultivated in the Oxford Botanic Garden since at least 1690. By 1800 it was an established alien on the walls in Oxford, and reached the Great Western Railway system in the city about 1879. Once established on the dry, sunny banks of the railway system its plumed seeds dispersed rapidly in the vortex of air following fast trains, and drifted in and out of carriages as the passengers entered and left. By the outbreak of the Second World War, it was well established and spreading over large parts of England and Wales. The progress of the dispersal of this ragwort has been thoroughly documented in a series of papers by Kent (1956–1964c). In Ireland, Praeger (1934) described the dispersal of *Senecio squalidus* along

Table 10.8 Common species in the railway flora

Arabidopsis thaliana (thale cress)	*Hypericum perforatum* (common St. John's wort)
Cardamine hirsuta (hairy bitter cress)	*Lathyrus sylvestris* (narrow-leaved everlasting pea)
Centaurea nigra (lesser knapweed)	*Leucanthemum vulgare* (ox-eye daisy)
Chamaenerion angustifolium (rose-bay willow-herb)	*Linaria vulgaris* (common toadflax)
Daucus carota (wild carrot)	*Lotus corniculatus* (birdsfoot-trefoil)
Equisetum arvense (common horsetail)	*Potentilla reptans* (creeping cinquefoil)
Erophila verna (spring whitlow grass)	*Senecio viscosus* (sticky groundsel)
Festuca rubra (red fescue)	*Tussilago farfara* (coltsfoot)
Fragaria vesca (wild strawberry)	*Vicia cracca* (tufted vetch)
Heracleum sphondylium (cow parsnip, hogweed)	

Source: Modified from Sargent (1984)

(a)

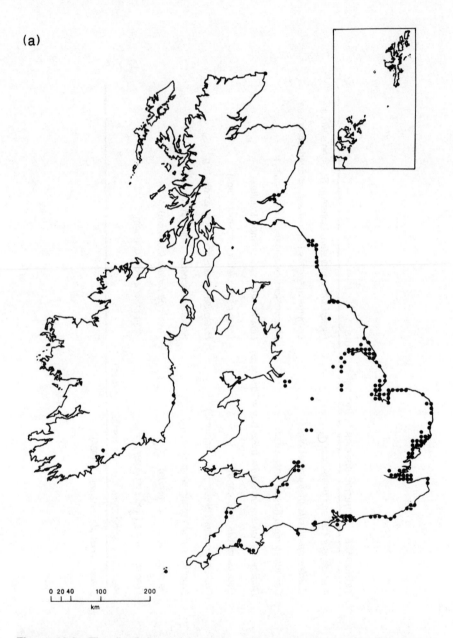

0 20 40   100   200
km

Figure 10.8   The distribution of the grass, reflexed poa (*Puccinellia distans*): (a) based on Perring and Walters (1962); (b) on roadsides, based on Scott and Davison (1982)

(b)

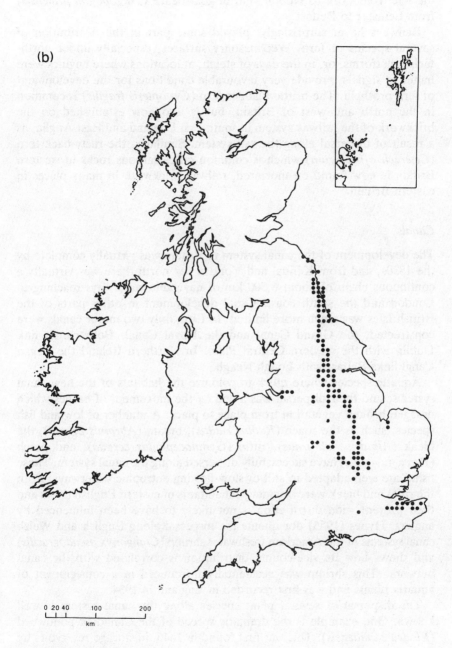

0 20 40   100   200
                km

Figure 10.8 continued

the line from Cork to Dublin, and of goatsbeard (*Tragopogon pratensis*) from Leinster to Belfast.

Railways have, surprisingly, played some part in the distribution of several species of fern. Wet masonry surfaces, especially under north-facing platforms, or, in the days of steam, at locations where engines were halted at signals, provide very favourable conditions for the development of fern prothalli. The brittle bladder-fern (*Cystopteris fragilis*) is common in the north and west of Britain, but is also now established on the brickwork of the railway system in south-east England and East Anglia, as a result of dispersal along the rail system. Similarly, the rusty-back fern (*Ceterach officinarum*) which is common on calcareous rocks in western Britain is now found on mortared, railway brickwork in many places in eastern Britain.

Canals

The development of the canal system in Britain was virtually complete by the 1830s, and from Kendal and York in the north there was virtually a continuous chain of about 6,500 km of navigable waterways reaching to London and the south coast. Canal development in other parts of the British Isles was much more limited. In Eire, only two major canals were constructed, the Grand Canal and the Royal Canal. Both canals link Dublin with the western Central Plain. In Northern Ireland the Lagan Canal links Belfast with Lough Neagh.

Aquatic species where quick to colonize the habitats of the new canal systems, and their dispersal was aided by the movement of barges which dragged broken vegetation from place to place. A number of lowland fish species, such as the roach (*Rutilus rutilus*), bream (*Abramis brama*), the bleak (*Alburnus alburnus*), ruffe (*Gymnocephalus cernua*), and perch (*Perca fluviatilis*) have successfully dispersed along the canal system. These fishes are well adapted to still or slow-flowing eutrophic waterways. Both the ruffe and bleak were originally inhabitants of eastern English rivers and their present wide distribution is not likely to have been influenced by anglers. Hynes (1955) documents the dispersal along English and Welsh canal systems of the American freshwater shrimp (*Crangonyx pseudogracilis*) and shows how its vice-county distribution is correlated with the canal network. This shrimp was accidentally introduced in a consignment of aquaria plants and was first recorded in England in 1934.

The dispersal of several plant species along the canal system is well known. One example is the dramatic spread of the Canadian pondweed (*Elodea canadensis*). This was first found in 1846, in storage reservoirs by Foxton Locks, on the canal near Market Harborough, Leicestershire. Since then it has spread throughout the inland waterway system. Praeger (1934) records how the canal system has extended the range of the fen

Table 10.9 The biota of contemporary London

	Number of species (seen in London)	Number of species (seen in the United Kingdom)	London species (as a percentage of United Kingdom species)
Higher plants	1,835	3,000	61
Insects			
Hemiptera/Homoptera (bugs)	317	390	82
Coleoptera (beetles)	248	3,700	7
Macro Lepidoptera (moths and butterflies)	728	930	78
Diptera (true flies)	2,300	5,200	44
Fishes (fresh and brackish water)	33	45	73
Amphibians	8	12	66
Reptiles	6	10	60
Birds	203	301	66
Terrestrial mammals			
Insectivora (shrews, moles, hedgehogs)	5	6	83
Chiroptera (bats)	10	15	66
Lagomorpha (hares and rabbits)	2	3	66
Rodentia (squirrels, voles, rats, mice)	8	17	47
Carnivora	4	11	36
Artiodactyla (deer)	3	9	33

Source: Modified from Gill and Bonnett (1984). London is defined as a circle of 32 km radius centred on St. Paul's Cathedral (about 3,200 km²)

pondweed (*Potamogeton coloratus*) eastward into Dublin County, and the opposite-leaved pondweed (*Groenlandia densa*) eastward into the heart of Dublin City.

Urban growth

Urban areas represent some of the most transformed habitats in the British Isles, but it would be wrong to think, as Elton did for his domestic habitats, that they are endless deserts of concrete and tarmac devoid of bio-geographical interest (Baines 1986). It is true that urban growth destroys habitats and displaces the flora and fauna. Soils are drained, their structures disturbed, and pavements, roads, and buildings seal off vast areas from sunlight, moisture, and gaseous exchange. On the other hand, new habitats are created, and, in spite of all the upheaval, one survey suggests that a surprising variety of plants and animals can persist (see Table 10.9).

The literature on urban biogeography in the British Isles is rather sparse, but this is slowly being put right as detailed urban mapping and recording schemes get underway, such as the West Midland Wildlife Survey (Teagle 1978). Much of the early impetus for studies of urban biogeography was as a result the profusion of wild flowers on bombed sites, created during the Second World War. As Fitter and Lousley (1953) note, the City of London had an impoverished flora and fauna before the war, but, as a result of the creation of blitzed areas, rapid colonization took place. For example, between 1939 and 1953, the City gained 269 plant, 3 mammal, 31 bird, and 56 insect species. The importance of urban wildlife is now recognized by the Nature Conservancy Council which, since 1983, has produced a quarterly newsletter entitled, *Urban Wildlife News*.

Most urban areas contain a wide variety of habitats, and much land is undeveloped (see Table 10.10). Of these urban habitats, Owen and Owen (1975) suggest that private gardens collectively constitute extremely important, but seriously neglected, nature reserves.

Table 10.10 Types of undeveloped land in urban areas

Institutional	Utilities	Domestic/Commercial
Schools and colleges	Sewage farms	Gardens/backyards
Hospitals/clinics	Reservoirs	Market gardens
Prisons	Canals	Industrial estates
Nature reserves	Airports	
Clubs	Churchyards	
Offices	Cemeteries	
Golf courses	Rights-of-way	
Race courses		
Zoological/botanic gardens		

Source: Modified from Gill and Bonnett (1973)

Owen and Owen note the incredible diversity of plant species in most gardens, in marked contrast to the monoculture of much of the country-side. They go on to suggest that the assumptions that gardens are barren, that imported plants harbour few animals, and that urban and suburban areas are an unmitigated ecological disaster, are very far from the truth. To put things right, they spent a number of years studying three groups of insects (butterflies, hoverflies, and ichneumonid wasps) that came into their 685 m^2 garden, situated on a corner plot on a busy road, about 7 km from the centre of Leicester. The results were remarkable. They showed that about a quarter of all British species of these three insect taxa were present at some time in their garden. Their message is clear: look and ye shall find.

One of the most thorough studies of the impact of urbanization in Great Britain on fauna is the Rothamsted Insect Survey (Taylor *et al*. 1978). In this study, a national network of 172 light traps was used to catch night-flying large moths (the Macrolepidoptera) over a period of at least a year, and in several cases much longer. The total trap catch was counted and individual species identified. To judge how rich the fauna was at each site, Taylor and his co-workers calculated an index of species diversity which is independent of the sample size (alpha diversity). The larger the value of alpha, the greater the species diversity. (Lewis and Taylor (1974) give details of its calculation.) They then went on to investigate how moth diversity was related to land use. This was achieved by calculating the area of various land uses within a 64.4 m radius of a sample site. Each land use was then given the following weightings: buildings = 0; grass and arable = 2; gardens = 3; woodland = 6; hedges = 29. A site index was estimated as the summation of the product of each land use and its weighting. It was then possible to model the relationship between moth diversity and the site index, using regression analysis. The following model was obtained:

alpha diversity = 25.94 + 0.034 (site index).

This model shows that as the land use changes from an agricultural to an urban character, the site index falls, producing a drop in species diversity. The model tells us that the alpha diversity for a trap site totally surrounded by buildings is 25.94, since the site index is zero.

Within any one urban area, we might expect a trend in species diversity along a transect from the edge to the centre, especially if the urban development is more or less concentrically zoned. Davis (1982) examined this hypothesis for ground-living arthropods in the City of London. He found that the number of species declined by nearly 60 per cent between the urban fringe and the city centre. He suggested four possible reasons for this trend: increasing disturbance, including the effects of pollution; a reduction in the total area of exploitable habitat with increasing urbanization (a species-area effect); a reduction in habitat diversity or variety of niches; some other urban-induced factor such as the heat-island effect.

The same trend in declining species richness toward the centre of the City of London was also found by Cousins (1982) in his study of snails and birds. Cousins also tested the hypothesis that as 'green'-patch size declined toward an urban centre then so, too, might the actual size of species. His analysis of data from the *Atlas of Breeding Birds of the London Area* (Montier 1977) shows that average species weight ranges from 122 g on London's outskirts to 90 g at the centre. He suggests that the size of an organism determines its food demand, and the size of prey it may eat. Each organism consumes 'packets' of food, and is itself a 'packet' of food to another organism. This may, he suggests, offer an explanation to the paradox that while bird species are, on average, small in towns, characteristic urban-species such as the feral pigeon (*Columbia livia*) and the herring gull (*Larus argentatus*) are quite large. This is because the 'packet' size of waste food produced by human activity – such as grain spills and household-waste food – is also relatively large and suitable for such species.

Several researchers make the point that there is a continuous supplementation to the urban biota from species in the surrounding countryside. Some rural–urban migrants are successful, some are not. A well-studied, successful migrant is the red fox (*Vulpes vulpes*) which has taken up residence in many British cities from Nottingham southward, and also in Glasgow and Edinburgh. A detailed questionnaire survey of inspectors of the Royal Society for the Prevention of Cruelty to Animals revealed a number of interesting regional variations in the colonization of urban and suburban areas by foxes (see Table 10.11) One noticeable finding was the large number of towns in the south-east of England with urban foxes, in comparison to the low incidence in the north. Furthermore, the questionnaire revealed that in six out of the seven regions studied, larger conurbations were more inclined to support foxes than were smaller.

MacDonald and Newdick (1982) suggest that there are differences in the habitat structures of large and small towns, which may underlie these trends. Certainly, we might expect fewer leafy suburbs in the industrial towns of the north of England, and there is reason to believe that working-class urban areas with large stray-dog populations are avoided by foxes. Harris (1981), for example, in his study of foxes in Bristol, found a very significant negative spatial correlation between foxes and stray dogs. In one neighbourhood, during a 23-day study, sixty-eight different stray dogs were recognized in an area of 0.2 km^2. The dogs roamed freely in the streets, gardens, and waste land – which would make it difficult for foxes to find a quite place to lie up during the day.

A particularly novel approach to the study of foxes within an urban area was that adopted by Harris and Rayner (1986). They made detailed surveys of fox distributions in Bath, Bournemouth, Poole, and the

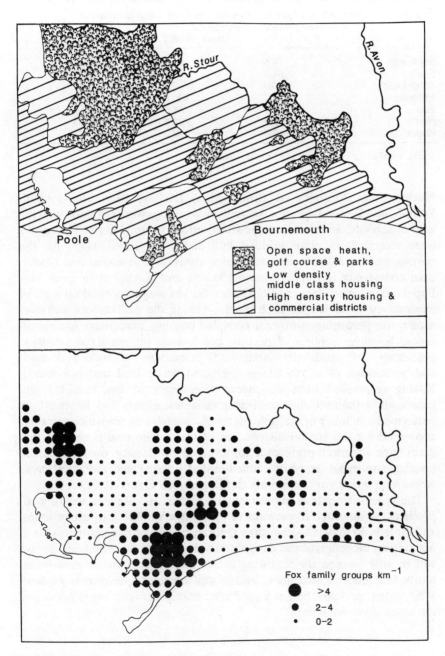

Figure 10.9 The distribution of fox families in the Bournemouth area in relation to land use. (Based partly on Harris and Rayner 1986)

Table 10.11 The distribution of urban foxes in England and Wales

Region	Percentage of towns with foxes present	Average town population (thousands) in survey with	
		foxes present	foxes absent
South-east	74.6	69.6	53.6
South	64.0	63.1	28.9
South-west	21.0	150.5	15.9
Midlands	52.7	136.9	37.3
East	31.4	66.6	53.5
North	15.9	129.3	78.5
Wales	47.8	12.9	61.3

Source: Modified from Macdonald and Newdick (1982)

Metropolitan Boroughs of Birmingham, Dudley, Sandwell, Solihull, Walsall, and Wolverhampton. Individual fox sitings were obtained with the help of schools, and the data assembled into fox family units. The results were recorded on maps gridded into 500 m^2 squares, which for the purposes of the subsequent statistical analysis were averaged into blocks, each comprising nine grid squares. Harris and Rayner then used 1981 Population Census data, and Ordnance Survey maps, to build up a set of explanatory variables for each block. Among the explanatory variables used were: percentage of owner-occupied housing; percentage of council-rented housing; number of persons per household; population density; percentage of residential land; and percentage of industrial land; and percentage of urban fringe (including waste land and woodland). Having assembled both dependent (frequency of fox families) and independent (habitat characteristics) variables, Harris and Rayner then performed a number of multiple regression analyses to see which variables accounted for the spatial distribution of foxes. The results showed that foxes have a marked preference for areas of low-density, owner-occupied housing and avoid areas of council-rented housing and industrial areas where vegetation cover is poorly developed (see Figure 10.9).

The types of analysis carried out by Harris and Rayner are of real, practical, significance if we want to encourage wildlife into urban areas, and there is no reason – other than ignorance – why urban planners should not employ biogeographical design principles. Let us hope that one day we will be able to read the following in an estate agent's house description blurb: 'all modern amenities, and a well-stocked, species-rich garden'. Who knows, perhaps there is a good relationship between house prices and the alpha diversity index!

Glossary

abiotic: the non-living factors affecting an ecosystem.

acid: applied to soils or soil water in which free acids occur. Such soils lack lime and other basic substances.

adventive: a poorly established alien species.

agamospermy: non-sexual production of seeds and embryos (literally, unmarried sex).

age and area theory: a theory that proposes that the greater the age of a taxon, the greater its geographical range.

alien: a species either deliberately or accidentally released into an area in which it has not occurred in historical time. Aliens may become 'naturalized' or remain 'casual' species.

alkaline: applied to soils or soil water rich in soluble bases such as lime or potash.

allogenic succession: succession due to external environmental changes.

allopatric: living in isolated geographical regions.

allopatric speciation: speciation due to geographical separation.

alpha diversity: the number of species within the community.

annual: a plant completing a life-cycle within a year.

apomixis: non-sexual reproduction in plants, including vegetative reproduction and the production of seeds and embryos (adj. apomictic).

Arctic: refers to regions within the Arctic Circle or to animals and plants found roughly in that area.

ASSI: area of special scientific interest (Eire).

autogenic succession: succession due to the actions of the organisms themselves which bring about changes in the environment, such as the shallowing of a lake due to organic-matter accumulation.

autotroph: an organism that uses sunlight to fix carbon dioxide and obtains other materials from inorganic sources.

bases: substances such as lime and potash which neutralize soils and soil water (adj. basic).

bathymetric sliding: the submergence of étages with decreasing latitude.

benthic: adjective applied to plants and animals living in, and on the sea-floor.

benthos: collective term for benthic organisms.

biennial: a plant with a life-cycle extending from one year to the next.

biochore: a plant climatic boundary.

biomass: literally the living weight of a component in an ecosystem, usually expressed as dry matter per unit area.

biota: The group of plants and/or animals that live together in one place or area.

biotic: pertaining to living organisms.

bog: an area of vegetation developed on nutrient-poor waterlogged ground, usually in areas of high precipitation.

boreal: (adj.) northern.

BP: before present. In radiocarbon dating, before AD 1950.

Britain: the main island of the British Isles, comprising the mainland of England, Scotland, and Wales.

British Isles: an archipelago of some 400 islands lying on the continental shelf off north-west Europe. Politically comprises the United Kingdom of Great Britain and Northern Ireland, the Irish Republic and the locally autonomous Isle of Man and Channel Islands.

brown earth: the characteristic soil type under deciduous forest in the British Isles.

calcareous: containing lime.

calcicole: naturally found in soils with a high lime-content.

calcifuge: naturally growing only in soils with low lime-content.

canonical distribution: a log–normal statistical distribution of species numbers; used in island biogeography.

cardinal temperatures: literally, the fundamental temperatures – the upper and lower lethal temperatures and the optimum temperature.

catadromous: a species that breeds in seawater but lives in fresh water, e.g. the eel.

Cenozoic: era of geological time comprising the Tertiary and Quaternary Periods.

choropleth map: a map in which statistical data are presented within fixed zones; sometimes referred to as a 'quantity in area' map. The statistical data may be simple presence/absence or grouped into classes.

chromosome: a thread-like structure in the cell nucleus composed of a backbone of protein surrounded by a double helix of DNA.

cladistic biogeography: the combination of vicariance biogeography with the science of cladistics.

cladistics: a method of describing the pathways of ancestry by means of a branching tree-like diagram called a cladogram.

cladogram: a tree-like diagram showing cladistic relationships. The vertical scale is usually time or evolutionary advancement.

climax: a relatively stable type of vegetation reached in succession. Used by some biogeographers to refer to a final form of vegetation reached under specific climatic conditions.

cline: a geographical gradient in a measurable characteristic such as species colour or temperature resistance.

Coleoptera: an order of beetles.

commensalism: an interspecific relationship of two organisms in which only one draws benefit and there is no apparent harm to the other.

compensation point: the point at which photosynthetic energy gains are equal to respiration energy losses, and the net production is zero.

competitive exclusion: the idea that no two species can occupy the same niche and coexist indefinitely.

continentality: a measure of likeness to a continental climate which is characterized by temperature extremes both in summer and winter.

corridor route: one which allows the spread of the majority of taxa from one region to another.

cryptogams: plants not reproducing by seeds.

Cybele: in Greek mythology, the Phrygian goddess of nature and fertility. A name chosen by H. C. Watson to describe a systematic description of the geography of plants of a specified area (see Flora).

deme: a local breeding sub-unit.

demersal: adjective applied to organisms living in deep water near the sea floor.

denizen: an apparently native plant but under suspicion of having been originally introduced.

Devensian glaciation: the last glacial period in Britain. Named from a type-site near Chester (Deva).

diapause: a period of suspended growth and reduced metabolism in the life-cycle of many insects.

diaspore: any stage, or part, of a plant that is adapted for dispersal.

diatom: a small unicellular alga with a cell wall impregnated with silica.

diploid: an organism having two haploid sets of chromosomes in the cell nuclei.

disjunct distribution: a discontinuous range in which two closely related populations are separated in geographic space.

disjunction: part of the geographic range of a species that is separate from the rest (see *allopatric*).

dispersal: the separation of an individual from other members of the population (see *migration*).

disseminule: any part of a plant that is used for dispersal, such as a spore, seed, or fruit.

ecology: the study of relationships between an organism and its environment.

ecosystem: an interacting and interdependent community of living organisms together with their abiotic environment.

ecotope: a particular portion of the physical world that forms the home for the organisms which inhabit it.

ecotype: a sub-species that is adapted to a special set of environmental conditions.

edaphic: relating to the soil.

element: a group of species occupying a characteristic geographical range which extends into the British Isles.

endemic: a species (or other taxa) limited in present range to a given geographical area.

epiphyte: a plant which grows on the surface of another but which is nutritionally independent (adj. epiphytic).

equilibrium theory: a theory in island biogeography which states that the number of species on an island is an equilibrium between immigration and extinction rates.

ESA: Environmentally Sensitive Area (Britain).

étage: communities of benthos with particular temperature requirements.

eury: a prefix, meaning narrow, used to describe the relative tolerance of a species with regard to an environmental factor.

eurytrophic: having a wide range of tolerance to both habitats and environmental conditions.

eutrophic: nutrient rich.

exclusive species: a plant which can only grow in a special habitat or environment.

Fauna: in Greek mythology, the goddess sister of Faunus. A term used to describe a systematic description of the animal families, genera and species of a region (plural – faunas, faunae).

fen: an area of vegetation adapted to waterlogged conditions, usually in a valley bottom where ground water supplies nutrients, often giving rise to a rich flora.

filter: a barrier that allows parts of fauna or flora to cross.

Flandrian: the present temperate stage which started about 10,000 BP. Synonymous with Holocene and Post-glacial.

Flora: the Roman goddess of flowers. A term used to describe a systematic description of plant families, genera, and species living in a particular area (plural – florae, floras).

flush: an area of soil enriched by transported materials such as dissolved mineral salts.

founder effect: genetic alteration in a colonizing population due to the founder principle.

founder principle: the idea that isolated colonists (founders) contain only a part of the ancestral gene pool, and so may evolve differently.

fugitive: species, often short-lived, which rapidly colonize newly exposed or disturbed habitats.

gamete: a reproductive cell containing a haploid number of chromosomes.

gene: a unit of the DNA (deoxyribonucleic acid) molecule coded to produce a specific protein.

generalized track: several lines on a map connecting two or more areas of endemism.

genotype: the genetic make-up of an organism (adj. genotypic).

genus: a group of allied species (pl. genera).

gley: a permanently waterlogged mineral soil characterized by blue-grey mottling in the upper horizons.

GPP: gross primary productivity.

grid: an arbitrary rectangular mesh drawn on a map and used for location definition instead of latitude and longitude.

habitat: literally, the place where an organism lives.

haploid: having the number of chromosome characteristic of a gamete.

herbivore: an organism that feeds chiefly on plant food.

heterotroph: an organism that obtains its energy and materials from other organisms.

homeothermic: pertaining to 'warm-blooded' animals that can regulate their body temperature.

indigenous: a species present in an area before man or having spread naturally into an area after the arrival of man (noun, indigene).

interstadial: a minor climatic amelioration during a cold stage.

ion: an electrically charged atom or molecule. Cations are positively charged with a deficit of electrons; anions are negatively charged with a surplus of electrons.

ITE: Institute of Terrestrial Ecology (Britain).

J: Joule, a unit of energy. 1 calorie = 4.187 J.

kJ: kilojoule (1,000 joules).

K-selection: selection favouring a more efficient utilization of resources.

lapse rate: the rate of decrease in air temperature which normally occurs with increasing altitude.

life-form: a classification of plant growth types according to the position of their perennating organs (e.g. buds) – developed by Raunkiaer.

MAFF: Ministry of Agriculture, Fisheries, and Food (Britain).

map projection: a method by which part or the whole of the earth's surface can be represented on a plane surface.

274

mesophyte: a plant living under neither very wet nor very dry conditions.

migration: the movement of an individual or group of organisms in relation to geographical co-ordinates.

MJ: Megajoule (1 million joules).

monoclimax: a theory that states that a single climax vegetation will develop under one climatic regime.

NCC: Nature Conservancy Council (Britain).

neritic: the shallow-water province of the pelagic environment. In the British seas it includes all waters over the continental shelf.

niche: the sum total of the various ways in which a species conforms to its particular environment.

NPP: net primary productivity. The rate at which plants and animals store energy in excess of that used in respiration. Usually expressed in dry weight of organic matter per unit area.

nunatak: high ground not inundated by a glacier.

oceanicity: as applied to climate to suggest equability due to the moderating influence of the surrounding waters.

oligotrophic: nutrient poor (see *eutrophic*).

ontogenetic: a migration circuit completed only once in a lifetime as in the case of the eel and salmon.

palynology: the study of fossil pollen.

pelagic: the marine organisms which are either active swimmers (nekton) or passively drifting species (plankton).

perennating bud: the overwintering, undeveloped bud of a perennial plant.

perennial: a plant living more than 3 years.

pH: a measure of the concentration of hydrogen ions in a solution: defined as $-\log_{10}H^+$.

phenotype: the outer appearance of an organism (adj. phenotypic).

plagioclimax: a climax community developed as a result of man-made influences (Greek, plagios = oblique or deflected).

plagiosere: a succession deflected from its normal outcome by man's influences.

plankton: minute plants and animals which float near the surface of fresh and salt water.

plate tectonics theory: a theory which suggests that the earth's crust is fractured into a number of rigid plates which move relative to each other. Continental drift is one consequence of plate tectonic movements.

podsol: a leached soil characteristic of areas of high rainfall and acid rocks. Typically the upper mineral horizon is grey and ash-like.

poikilotherm: an organism whose body temperature is always close to the temperature of the medium in which it lives.

pollen: in seed-bearing plants, the grains containing the male gametes (sex cells).

propagule: any part of a species or stage in its life-cycle that can reproduce and establish a new population.

r-selection: selection favouring higher population-growth rates and higher productivity.

race: members of partially isolated demes.

range: the area normally occupied by a breeding population of a species.

relict: species now restricted to favourable sites whose distribution was formerly more widespread.

RSPB: Royal Society for the Protection of Birds.
ruderal: plants tolerant of highly disturbed ground and waste places.

sere: the sequence of identifiable stages in the development of an ecosystem (adj. seral).
speciation: the process by which two or more contemporaneous species evolve from a single ancestral population.
species: a population, the individuals of which are actually or potentially capable of breeding to produce fertile offspring.
species-area curve: a log–log plot relating number of species of a given taxon, for example: butterflies and island area; true islands or surrogate islands, such as parks in a town.
species pool: in island biogeography, the number of species able to immigrate to an island from the mainland.
spp.: an abbreviation for sub-species.
SSSI: Site of Special Scientific Interest (Britain).
stadial: a period of ice advance during a cold stage.
standing crop: equivalent to the term biomass (also called standing stock).
steno: a prefix, meaning narrow, used to describe the relative tolerance of an organism with regard to some factor.
stenophagous: a species which is very restricted as to its choice of food. In insects may be a single species of plant.
sub-species: a population partially fertile with its original parent species.
sweepstakes dispersal: low probability dispersal across a formidable barrier.
symbiosis: a relationship between two unrelated species each of which receives some benefit.
sympatric speciation: Speciation within the same geographical area due to chromosome mutation or habitat selection.

taxon: a useful general term for any taxonomic category e.g. genus or species (plural, taxa).
taxonomy: the science of classification.
tetrad: a two-kilometre grid square used as the basic mapping unit for many local mapping schemes. Each 10 km grid square contains 25 tetrads.
texture: the dominant particle size of the soil mineral component as measured in the fine-earth fraction (less than 2 mm in diameter).
thermophilous species: warmth-demanding species.
translocation: the vertical movement of soluble and insoluble substances in a soil profile.
trophic level: the grouping of organisms in an ecosystem according to their food sources.

UTM: Universal Transverse Mercator projection.

variety: a recognizable type within a deme.
vicariance biogeography: the study of the historical evolution of distributions through an examination of taxonomic relationships.
vicariant: an adjective describing taxonomically related varieties, species, or genera having distinct non-overlapping ranges (noun – vicariance).
vicariist: someone who studies biogeography within a vicariance paradigm.
vice-county: a unit area originally devised by H. C. Watson for the purpose of recording species distributions.

waif dispersal: long-distance dispersal across large barriers.
weed: a plant considered undesirable by man.
wildwood: a site continuously wooded since the climatic optimum.

zoogeography: the study of the distribution of animals.

Bibliography

Alford, D. A. (ed.) (1973) *Provisional Atlas of the Insects of the British Isles, Part 3* Hymenoptera – Apidae, *Bumblebees*, Abbots Ripton: Biological Records Centre.

Alphey, T. J. W. and Taylor, E. (1986) *European Atlas of the* Longidoridae *and* Trichodoridae, Dundee: Scottish Crop Research Institute.

Anderson, P. and Yalden, D. W. (1981) 'Increased sheep numbers and the loss of heather moorland in the Peak District, England', *Biological Conservation*, vol. 20, pp. 195–213.

Babington, C. C. (1859) 'Hints towards a *Cybele Hibernica*', *Natural History Review*, vol. 6, pp. 533–7.

Baines, C. (1986) *The Wild Side of Town*, London: Elm Tree Books.

Baker, R. R. (1982) *Migration Paths Through Time and Space*, London: Hodder & Stoughton.

Bannister, P. (1976) *Introduction to Physiological Plant Ecology*, Oxford: Blackwell.

Barr, C., Benefield, C., Bunce, R., Ridsdale, H., and Whittaker, M. (1986) *Landscape Changes in Britain*, Grange-Over-Sands: Institute of Terrestrial Ecology.

Bary, B. McK. (1963) 'Distributions of Atlantic pelagic organisms in relation to surface water bodies', in M. J. Dunbar (ed.) *Marine Distributions*, Royal Society of Canada Special Publications no. 5, Toronto: University of Toronto Press. pp. 51–67.

Beaufort, L. F. de (1951) *Zoogeography of the Land and Inland Waters*, London: Sidgwick & Jackson.

Begon, M. and Mortimer, M. (1981) *Population Ecology*, London: Blackwell.

Beirne, B. P. (1952) *The Origin and History of British Fauna*, London: Methuen.

Bell, F. G. (1970) 'Late Pleistocene floras from Earith, Huntingdon', *Philosophical Transactions of the Royal Society, B*, vol. 258, pp. 347–78.

—— and Dickson, C. A. (1971) 'The Barnwell Station arctic flora: a reappraisal of some plant identifications', *New Phytologist*, vol. 70, pp. 627–36.

Berry, R. J. (1977) *Inheritance and Natural History*, London: Collins.

Birks, H. J. B. (1986) 'Late-Quaternary biotic changes in terrestrial and lacustrine environments, with particular reference to north-west Europe', in B. E. Berglund (ed.) *Handbook of Holocene Palaeoecology and Palaeohydrology*, Chichester: Wiley. pp. 3–65.

Birks, H. J. B. and Birks, H. H. (1980) *Quaternary Palaeoecology*, London: Edward Arnold.

Birks, H. J. B. and Deacon, J. (1973) 'A numerical analysis of the past and present flora of the British Isles', *New Phytologist*, vol. 72, pp. 877–902.

Blackburn, K. B. (1931) 'The Late-Glacial and Post-Glacial periods in the north

Pennines. II. Possible survivals in our flora', *Transactions of the Northern Naturalist's Union*, vol. 1, pp. 30–6.

Boulter, M. C. (1971) 'A palynological study of two Neogene plant beds in Derbyshire', *Bulletin, British Museum of Natural History (Geology)*, vol. 19, pp. 361–410.

—— (1980) 'Irish Tertiary plant fossils in a European context'. *Journal of Earth Sciences Royal Dublin Society*, vol. 3, pp. 1–11.

Bowen, D. Q. (1978) *Quaternary Geology*, Oxford: Pergamon.

Bowen, D. Q., Rose, J., McCabe, A. M., and Sutherland, D. G. (1986) Correlation of Quaternary glaciations in England, Ireland, Scotland and Wales, *Quaternary Science Reviews*, vol. 5, pp. 299–341.

Bridges, E. M. and Davidson, D. A. (1982) *Principles and Applications of Soil Geography*, London: Longman.

Briggs, J. C. (1974) *Marine Zoogeography*, New York: McGraw-Hill.

Brown, A. H. F. (1974) 'Nutrient cycles in oakwood ecosystems in N. W. England', in M. G. Morris and F. H. Perring (eds) *The British Oak*, Faringdon: Classey, pp. 141–61.

Brown, J. H. and Gibson, A. C. (1983) *Biogeography*, St Louis: Mosby.

Browne, J. (1983) *The Secular Arc*, New Haven: Yale University Press.

Burnham, C. P. (1970) 'The regional pattern of soil formation in Great Britain', *Scottish Geographical Magazine*, vol. 86, pp. 25–34.

—— and Mackney, D. (1964) 'Soils of Shropshire', *Field Studies*, vol. 2, pp. 83–113.

Cabot, D. (ed.) (1985) *The State of the Environment*, Dublin: An Foras Forbartha.

Cadbury, C. J. (1987) 'UK estuaries under threat'. *RSPB Conservation Review*, no. 1, pp. 41–7.

Cadbury, D. A., Hawkes, J. G., and Readett, R. C. (1971) *A Computer-Mapped Flora of Warwickshire*, London: Academic Press.

Carson, R. (1963) *Silent Spring*, New York: Hamilton.

Carter, R. N. and Prince, S. D. (1985) 'The effect of climate on plant distributions', in M. J. Tooley and G. M. Sheail (eds) *The Climatic Scene*, London: George Allen & Unwin, pp. 235–54.

Chandler, M. E. J. (1921) 'The arctic flora of the Cam Valley at Barnwell, Cambridge'. *Quarterly Journal of the Geological Society of London*, vol. 77, pp. 4–22.

Charlesworth, J. K. (1930) 'Some geological observations on the origin of the Irish fauna and flora', *Proceedings of the Royal Irish Academy (B)*, vol. 39, pp. 358–90.

Clapham, A. R. (ed.) (1978) *Upper Teesdale. The Area and Its Natural History*, London: Collins.

Clements, F. E. (1916) *Plant Succession: an Analysis of the Development of Vegetation*, Washington D.C.: Carnegie Institution.

Colebrook, J. M. *et al.* (1961) 'Continuous plankton records: contributions towards a plankton atlas of the north-eastern Atlantic and the North Sea', *Bulletin of Marine Ecology*, vol. 5, pp. 65–111.

Collingbourne, R. H. (1976) 'Radiation and sunshine', in T. J. Chandler and S. Gregory, *The Climate of the British Isles*, London: Longman, pp. 74–95.

Common, R (1970) 'Land drainage and water use in Ireland', in N. Stephens and R. E. Glassock, *Irish Geographical Studies. In honour of Estyn Evans*. Belfast: Queen's University, pp. 342–60.

Condry, W. M. (1981) *The Natural History of Wales*, London: Collins.

Conolly, A. P. and Dahl, E. (1970) 'Maximum summer temperature in relation to the modern and Quaternary distributions of certain arctic-montane species in the British Isles', in D. Walker and R. West (eds), *Studies in the Vegetation History of the British Isles*, Cambridge: Cambridge University Press, pp. 159–223.

Coope, G. R. (1959) 'A late Pleistocene insect fauna from Chelford, Cheshire', *Proceedings of the Royal Society (B)* vol. 151, pp. 70–86.

Coope, G. R. (1970) 'Climatic interpretations of late Weichselian *Coleoptera* from the British Isles'. *Révue de Géographie Physique et de Géologie Dynamique*, vol. 12, pp. 149–55.

—— (1977) 'Fossil coleopteran assemblages as sensitive indicators of climatic changes during the Devensian (last) cold stage', *Philosophical Proceedings of the Royal Society, B,* vol. 280, pp. 313–37.

—— (1979) 'Late Cenozoic fossil Coleoptera: evolution, biogeography, and ecology', *Annual Review of Ecology and Systematics*, vol. 10, pp. 247–67.

——, Shotton, F. W. and Strachan, I. (1961) 'A late Pleistocene fauna and flora from Upton Warren, Worcestershire', *Philosophical Proceedings of the Royal Society, B,* vol. 244, pp. 379–421.

Corbet, G. B. (1962) 'The Lusitanian element in the British fauna', *Science Progress*, vol. 50, pp. 177–91.

Countryside Commission (1987) *Protected Areas in the United Kingdom*, Cheltenham: Countryside Commission.

Cousins, S. H. (1982) 'Species size distributions of birds and snails in an urban area', in R. Bornkamm, J. A. Lee, and M. R. D. Seaward (eds) *Urban Ecology*, Oxford: Blackwell, pp. 99–109.

Cox, C. B. and Moore, P. D. (1985) *Biogeography. An Ecological and Evolutionary Approach* (fourth edition), Oxford: Blackwell.

Crisp, D. J. (1958) 'The spread of *Elminius modestus* Darwin in north-west Europe', *Journal of the Marine Biological Association, U K*, vol. 37, pp. 483–520.

—— and Southwood, A. J. (1958) 'The distribution of intertidal organisms along the coasts of the English Channel', *Journal of the Marine Biological Association, UK*, vol. 37, pp. 157–208.

Croizat, L. (1958) *Panbiogeography*, 2 volumes, Caracus: The Author.

Cruickshank, J. G. (1984) 'Soils and biogeography in Ireland, 1934–1984', *Irish Geography*, vol. 17A, pp. 97–116.

Curtis, L. F., Courtney, F. M., and Trudgill, S. (1976) *Soils in the British Isles*, London: Longman.

Cushing, D. H. (1975) *Marine Ecology and Fisheries*, Cambridge: Cambridge University Press.

Dandy, J. E. (1969) *Watsonian Vice-Counties of Great Britain*, London: The Ray Society.

Darlington, P. J. (1959) 'Darwin and zoogeography', *Proceedings of the American Philosophical Society*, vol. 103, pp. 307–19.

Darwin, C. (1859) *On the Origin of Species by means of Natural Selection, or the Preservation of Favoured Races in the Struggle for Life*, London: Watts.

Davies, J. L. (1957) 'The geography of the gray seal', *Journal of Mammalogy*, vol. 38, pp. 297–310.

Davis, B. N. K. (ed.) (1981) *The Ecology of Quarries*, Abbots Ripton: Institute of Terrestrial Ecology.

—— (1982) 'Habitat diversity and invertebrates in urban areas', in R. Bornkamm, J. A. Lee, and M. R. D. Seaward (eds) *Urban Ecology*, Oxford: Blackwell, pp. 49–65.

Deville, J. St Clair (1930) 'Quelques aspects du peuplement des Iles britanniques. (Coléoptères)', *Mémoires de la Société de Biogéographie*, vol. 3, pp. 99–150.

Devoy, R. J. (1985) 'The problem of Late Quaternary land bridges between Britain and Ireland', *Quaternary Science Reviews*, vol. 4, pp. 43–58.

Dickson, J. H. (1973) *Bryophytes of the Pleistocene*, Cambridge: Cambridge University Press.

Dodge, J. (ed.) (1981) *Provisional Atlas of the Marine Dinoflagellates of the British Isles*, Abbots Ripton: Institute of Terrestrial Ecology.

Dony, J. G. (1955) 'Notes on the Bedfordshire railway flora', *Bedfordshire Naturalist*, vol. 9, pp. 12–6.

—— (1968) 'The B.S.B.I. in a changing Britain', *Proceedings of the Botanical Society of the British Isles*, vol. 7, pp. 311–23.

—— (1976) *Bedfordshire Plant Atlas*, Luton: Luton Museum and Art Gallery.

Douglas, I. (1981) 'The city as an ecosystem', *Progress in Physical Geography*, vol. 5, pp. 315–67.

—— (1983) *The Urban Environment*, London: Edward Arnold.

Dring, M. J. (1982) *The Biology of Marine Plants*, London: Edward Arnold.

Drury, W. H. and Nisbet, I. C. T (1973) 'Succession', *Journal of the Arnold Arboretum (Harvard University)*, vol. 54, pp. 331–68.

Duff, A. G. and Lowe, P. D. (1981) 'Great Britain', in E. J. Kormondy and J. F. McCormick, (eds) *Handbook of Contemporary Developments in World Ecology*, Westport, Conn.: Greenwood Press, pp. 141–56.

Earll, R. and Farnham, W. (1983) 'Biogeography', in R. Earll and D. G. Erwin, (eds) *Sublittoral Ecology – the Ecology of the Shallow Sublittoral Benthos*, Oxford: Clarendon Press. pp. 165–209.

Edwards, K. J. and Warren, W. P. (1985) *The Quaternary History of Ireland*, London: Academic Press.

Egerton, F. N. (1979) 'Hewett C. Watson, Great Britain's first phytogeographer', *Huntia*, vol. 3, pp. 87–102.

Ekman. S. (1953) *Zoogeography of the Sea*, London: Sidgwick & Jackson.

Elkington, T. T. (1984) 'Cytogenetic variations in the British flora: origins and significance', *New Phytologist*, vol. 98, pp. 101–18.

Elton, C. S. and Miller, R. S. (1954) 'The ecological survey of animal communities: with a practical system of classifying habitats by structural characters', *Journal of Ecology*, vol. 42, pp. 460–96.

Europa Publications (1988) *The Europa Year Book*, London: Europa Publications.

Evans, J. G. (1975) *The Environment of Early Man in the British Isles,* London: Elek.

Eyre, S. R. (1968) *Vegetation and Soils*, 2nd edition, London: Edward Arnold.

Fairbairn, W. A. (1968) 'Climatic zonation in the British Isles', *Forestry*, vol. 41, pp. 117–30.

Fairley, J. (1984) *An Irish Beast Book*, Belfast: Blackstaff Press.

Fenton, A. F. (1964) 'Atmospheric pollution of Belfast and its relationship to the lichen flora', *Irish Naturalist*, vol. 14, pp. 237–45.

Fichman, M. (1977) 'Wallace: zoogeography and the problem of land bridges', *Journal of the History of Biology*, vol. 10, pp. 45–63.

Fitter, R. S. R., and Lousley, J. E. (1953) *The Natural History of the City*, London: The Corporation of London.

Forbes, E. (1846) 'On the connexion between the distribution of the existing fauna and flora of the British Isles, and the geological changes which have affected their area, especially during the epoch of the northern drift', *Memoirs of the Geological Survey of Great Britain*, vol. 1, pp. 336–432.

—— (1859) *Outlines of the Natural History of Europe. The Natural History of the European Seas*. (edited and continued by R. Godwin-Austin). London: John van Voorst (reprinted in 1977, New York: Arno Press)

Ford, M. J. (1982) *The Changing Climate*, London: George Allen & Unwin.

Forman, R. T. T. and Baudry, J. (1984) 'Hedgerows and hedgerow networks in landscape ecology', *Environmental Management*, vol. 8, pp. 495–510.

Forsyth, J, and Buchanan, R. H. (eds) (1982) *The Ulster Countryside in the 1980s*, Belfast: Institute of Irish Studies, Queens University.

Fosberg, F. R. (1976) 'Geography, ecology and biogeography', *Annals of the Association of American Geographers*, vol. 66, pp. 117–28.

Fryer, G. (1981) 'The copepod *Salmincola edwardsii* as a parasite of *Salvelinus alpinus* in Britain, and a consideration of the so-called relict fauna of Ennerdale Water', *Journal of the Zoological Society of London*, vol. 193, pp. 253–68.

Fuller, R. J. (1982) *Bird Habitats in Britain*, Calton: Poyser.

Furley, P. A. and Newey, W. W. (1983) *Geography of the Biosphere*, London: Butterworths.

Gadgil, M. and Solbrig, O. T. (1972) 'The concept of r- and K-selection: evidence from wild flowers and some theoretical considerations', *The American Naturalist*, vol. 106, pp. 14–31.

Gemmell, R. P. (1977) *Colonisation of Industrial Wasteland*, London: Edward Arnold.

George, M. (1977) 'The decline in Broadland's aquatic fauna and flora: a review of the present position', *Transactions of the Norfolk and Norwich Naturalists' Society*, vol. 24, pp. 41–53.

George, W. (1962) *Animal Geography*, London: Heinemann.

—— (1964) *Biologist Philosopher*, London: Abelard-Schuman.

Ghiselin, M. T. (1969) *The Triumph of the Darwinian Method*, Berkeley: University of California Press.

Gilbert, O. L. (1970) 'Further studies on the effect of sulphur dioxide on lichens and bryophytes', *New Phytologist*, vol. 69, pp. 605–27.

—— (1974) 'Air pollution survey by school children', *Environmental Pollution*, vol. 6, pp. 175–80.

Gill, D. and Bonnett, P. (1973) *Nature in the Urban Landscape: A Study of City Ecosystems*, Baltimore: York Press.

Gleason, H. A. (1924) 'Age and area from the viewpoint of phytogeography', *American Journal of Botany*, vol. 11, pp. 541–46.

Glemarec, M. (1973) 'The benthic communities of the European North Atlantic continental shelf', *Oceanography and Marine Biology: Annual Review*, vol. 11, pp. 263–89.

Godwin, H. (1940) 'Pollen analysis and forest history of England and Wales', *New Phytologist*, vol. 39, pp. 370–400.

—— (1956) *The History of the British Flora: A Factual Basis for Phytogeography*, Cambridge: Cambridge University Press.

—— (1967) 'The ancient cultivation of Hemp', *Antiquity*, vol. 41, pp. 42–50.

—— (1975) 'History of the natural forests of Britain: establishment, dominance and destruction', *Philosophical Transactions of the Royal Society of London, B*, vol. 271, pp. 47–67.

Goldie, H. (1986) 'Human influence on landforms: the case of limestone pavements', in K. Peterson and M. M. Sweeting (eds) *New Directions in Karst*, Norwich: Geo Books, pp. 515–41.

Good, R. (1930) 'The geography of the genus *Coriaria*', *New Phytologist*, vol. 29, pp. 170–98.

—— (1948) *A Geographical Handbook of the Dorset Flora*, Dorchester: Dorset Natural History Society.

—— (1974) *The Geography of Flowering Plants*, 4th edition (first published in 1947). London: Longman.

Goodman, T. and Roberts, T. M. (1971) 'Plants and soils as indicators of metals in the air', *Nature*, vol. 231, pp. 287–92.

Gorman, M. (1979) *Island Ecology*, London: Chapman and Hall.

Green, B. (1981) *Countryside Conservation*, London: George Allen & Unwin.

Greenwood, E. F. and Gemmell, R. P. (1978) 'Derelict land as a habitat for rare plants: s. Lancs (v.c. 59) and w. Lancs (v.c. 60)', *Watsonia*, vol. 12, pp. 33–40.

Grime, J. F. (1979) *Plant Strategies and Vegetation Processes*, Chichester: Wiley.

Gulliver, R. (1980) 'Once the trains have gone', *Country Life*, January, pp. 294–5.

Hagen, J. B. (1986) 'Ecologists and taxonomists: divergent traditions in twentieth-century plant geography', *Journal of the History of Biology*, vol. 19, pp. 197–214.

Harding, P. T. (1985) *Current Atlases of the Flora and Fauna of the British Isles 1985*, Abbots Ripton: Institute of Terrestrial Ecology.

—— and Sutton, S. L. (eds) (1985) *Woodlice in Britain and Ireland: Distribution and Habitat*, Abbots Ripton: Institute of Terrestrial Ecology.

Harding, R. J. (1978) 'The variation of the altitudinal gradient of temperature within the British Isles', *Geografiska Annaler*, vol. 60A, pp. 43–9.

—— (1979) 'Radiation in the British uplands', *Journal of Applied Ecology*, vol. 16, pp. 161–70.

Harley, J. B. (1975) *Ordnance Survey Maps. A Descriptive Manual*, London: HMSO.

Harris, S. (1981) 'An estimate of the numbers of foxes (*Vulpes vulpes*) in the city of Bristol, and some possible factors affecting their distribution', *Journal of Applied Ecology*, vol. 18, pp. 455–65.

—— and Rayner, J. M. V. (1986) 'Urban foxes (*Vulpes vulpes*): population estimates and habitat requirements in several British cities', *Journal of Animal Ecology*, vol. 55, pp. 575–91.

Harris, T. M. (1950) 'A great Piocene flora (review of W. Szafer, 1946–7)', *New Phytologist*, vol. 49, p. 421.

Harvey, H. W. (1950) 'On the production of living material in the sea off Plymouth', *Journal of the Marine Biological Association*, vol. 29, pp. 97–137.

Hawksworth, D. L. (ed.) (1974) *The Changing Flora and Fauna of Britain*, The Systematics Association, special volume no. 6, London: Academic Press.

—— and Rose, F. (1976) *Lichens as Pollution Monitors*, London: Edward Arnold.

Hayward, I. M. and Druce, G. C. (1919) *The Adventive Flora of Tweedside*, Arbroath: Buncle.

Healey, M. J. and Ilbery, B. W. (1985) *The Industrialization of the Countryside*, Norwich: Geo Books.

Heath, J. and Perring, F. (1978) *Biological Records Centre*, Abbots Ripton: Institute of Terrestrial Ecology.

——, Pollard, E. and Thomas, J. A. (1984) *Atlas of Butterflies in Britain and Ireland*, Harmondsworth: Viking.

Heslop-Harrison, J. (1953) 'The North American and Lusitanian elements in the flora of the British Isles', in J. E. Lousley (ed.) *The Changing Flora of Britain*, Arbroath: Buncle. pp. 105–23.

Hiscock, K. and Mitchell, R. (1980) 'The description and classification of sublittoral epibenthic ecosystems', in J. H. Price, D. E. G. Irvine, and W. F. Farnham, *The Shore Environment: Methods and Ecosystems, vol. 2*, London: Academic Press, pp. 323–69.

Hoch, E. (1982) 'Fossil evidence of early Tertiary North Atlantic events viewed in a European context', In M. P. H. Bott, S. Saxov, M. Talwani, and J. Thiede (eds) *Structure and Development of the Greenland–Scotland Ridge*, New York: Plenum, pp. 401–15.

Hoek, C. Van den (1982) 'The distribution of benthic marine algae in relation to the temperature regulation of their life histories', *Biological Journal of the Linnean Society*, vol. 18, pp. 81–144.
—— and Donze, M. (1967) 'Algal phytogeography of the European coasts', *Blumea*, vol. 15, pp. 63–89.
Hooker, J. D. (1834) *Geography Considered in Relation to the Distribution of Plants*, Reprinted in L. Maiorca (ed.) (1977) *Ecological Phytogeography in the Nineteenth Century*, New York: Arno Press.
Hooper, M. D. (1970a) 'The size and surroundings of nature reserves', *Symposium of the British Ecological Society*, vol. 11, pp. 555–61.
—— (1970b) 'The botanical importance of our hedgerows', *Botanical Society of the British Isles*, report 11, pp. 58–62.
Humphries, C. J. and Parenti, L. R. (1986) *Cladistic Biogeography*, Oxford: Clarendon Press.
Huntley, B and Birks, H. J. B. (1983) *An Atlas of Past and Present Pollen Maps for Europe: 0–13000 years ago*, Cambridge: Cambridge University Press.
Hutchins, L. W. (1947) 'The basis for temperature zonation in geographical distribution', *Ecological Monographs*, vol. 17, pp. 325–35.
Hutchinson, G. E. (1958) 'Concluding remarks', *Cold Spring Harbor Symposium on Quantitative Biology*, vol. 22, pp. 415–27.
Hynes, H. B. N. (1955) 'Distribution of some freshwater Amphipoda in Britain'. *Verhandlungen der Internationalen Vereinigung für theoretische und angerwandte Limnologie*, vol. 12, pp. 620–28.
—— (1978) *The Biology of Polluted Waters*, Liverpool: Liverpool University Press.
Institute of Terrestrial Ecology (1978) *Overlays of Environmental and other Factors for use with Biological Records Centre Distribution Maps*, Cambridge: Institute of Terrestrial Ecology.
Iversen, J. (1941) 'Land occupation in Denmark's stone age', *Danmarks Geologiske undersogelse, II*, vol. 66, pp. 1–68.
—— (1947) '*Viscum, Hedera* and *Ilex* as climatic indicators', *Geologiska Foreningen i Stockholm Fordhandlingar*, vol. 66, pp. 463–83.
—— (1958) 'The bearing of glacial and interglacial epochs on the formation and extinction of plant taxa', *Uppsala Universitets Arsskrift*, vol. 6, pp. 210–15.
Jalas, J. and Suominen, J. (eds) (1983) *Atlas Florae Europaeae: Distribution of Vascular Plants in Europe*, vol. 6, *Caryophyllaceae*, Helsinki: Societas Biologica Fennica Vanamo.
Jardine, N. (1972) 'Computational methods in the study of plant distributions'. in D. H. Valentine (ed.) *Taxonomy, Phytogeography and Evolution*, London: Academic Press, pp. 381–97.
Jelgersma, S.(1979) 'Sea-level changes in the North Sea Basin', in E. Oele, R. T. E. Schuttenhelm, and A. J. Wiggers (eds) *The Quaternary History of the North Sea*, Acta Universitatis Upsaliensis Symposia Universitatis Upsaliensis Annum Quingentesimum Celebrantis, 2, pp. 233–48.
Jermy, A. C. and Crabbe, J. A. (eds) (1978) *The Island of Mull, A Survey of its Flora and Environment*, London: British Museum (Natural History).
Kelsey, J. G. (1975) 'Industrial development and wildlife conservation', *Environmental Conservation*, vol. 2, pp. 99–108.
Kent, D. H. (1956) '*Senecio squalidus* L. in the British Isles – 1. Early records (to 1877)', *Proceedings of the Botanical Society of the British Isles*, vol. 2, pp. 115–18.
—— (1957) '*Senecio squalidus* L. in the British Isles – 3. East Anglia', *Transactions of the Norfolk and Norwich Naturalist Society*, vol. 18, pp. 30–1.
—— (1960) '*Senecio squalidus* L. in the British Isles – 2. The spread from Oxford

(1879–1939)', *Proceedings of the Botanical Society of the British Isles*, vol. 3, pp. 375–9.

—— (1964a) '*Senecio squalidus* L. in the British Isles – 4. Southern England (1940→)', *Proceedings of the Botanical Society of the British Isles*, vol. 5, pp. 210–13.

—— (1964b) '*Senecio squalidus* L. in the British Isles – 5. The Midlands (1940→)', *Proceedings of the Botanical Society of the British Isles*, vol. 5, pp. 214–16.

—— (1964c) '*Senecio squalidus* L. in the British Isles – 6. Northern England (1940→)', *Proceedings of the Botanical Society of the British Isles*, vol. 5, pp. 217–19.

Kerney, M. P. (ed.) (1976) *Atlas of the Non–Marine Mollusca of the British Isles*, Cambridge: Institute of Terrestrial Ecology.

Kinch, M. P. (1980) 'Geographical distribution and the origin of life: the development of early nineteenth century British Explanations', *Journal of the History of Biology*, vol. 13, pp. 91–119.

Kitching, J. A. and Ebling. F. J. (1967) 'Ecological studies at Lough Ine', *Advances in Ecological Research*, vol. 4, pp. 197–291.

Lack. D. (1969) 'Population changes in the land birds of a small island', *Journal of Animal Ecology*, vol. 38, pp. 211–18.

Larcher, W. (1975) *Physiological Plant Ecology*, Berlin: Springer-Verlag.

Lee, A. J. and Ramster, J. W. (eds) (1981) *Atlas of the Seas around the British Isles*, Lowestoft: MAFF Directorate of Fisheries Research.

Leith, H. and Box, E. (1972) 'Evapotranspiration and primary productivity', in J. R. Mather (ed.) *Papers on Selected Topics in Climatology*, New York: Elmer, pp. 37–44.

Lever, C. (1979) *The Naturalised Animals of the British Isles*, London: Granada.

Lewis, J. R. (1976) *The Ecology of Rocky Shores*, London: Hodder & Stoughton.

Lewis, T. and Taylor, L. R. (1974) *An Introduction to Experimental Ecology*, London: Academic Press.

Lousley, J. E. (1970) 'The influence of transport on a changing flora', in F. H. Perring (ed.) *The Flora of a Changing Britain*, Hampton: Classey, pp. 73–89.

Love, A. and Love, D. (1958) 'The American element in the flora of the British Isles', *Botaniska Notiser*, vol. III, pp. 376–88.

Macan, T. T. (1963) *Freshwater Ecology*, London: Longman.

MacArthur, R. H. and Wilson, E. O. (1963) 'An equilibrium theory of insular zoogeography', *Evolution*, vol. 17, pp. 373–87.

—— and —— (1967) *The Theory of Island Biogeography*, Princeton: Princeton University Press.

MacDonald, D. W. and Newdick, M. T. (1982) 'The distribution and ecology of foxes, *Vulpes vulpes* (L.) in urban areas', in R. Bornkamm, J. A. Lee, and M. R. D. Seaward (eds) *Urban Ecology*, Oxford: Blackwell, pp. 123–39.

McGuinness, K. A. (1984) 'Equations and explanations in the study of species-area curves', *Biological Reviews*, vol. 59, pp. 423–40.

McIntosh, R. P. (1985) *The Background of Ecology*, Cambridge: Cambridge University Press.

McKenna, M. C. (1975) 'Fossil mammals and early Eocene North Atlantic land continuity', *Annals of the Missouri Botanic Garden*, vol. 62, pp. 335–53.

MacNeill, N. (1968) 'Gridding', *Irish Naturalist*, vol. 16, pp. 73–4.

Marchant, C. J. (1967) 'Evolution in *Spartina* (Gramineae), 1. The history and morphology of the genus in Britain', *Journal of the Linnean Society, Botany*, vol. 60, pp. 1–24.

Matthews, J. R. (1937) 'Geographical relationships of the British flora', *Journal of Ecology*, vol. 35, pp. 1–90.

—— (1955) *Origin and Distribution of the British Flora*, London: Hutchinson.

Mellanby, K. (1967) *Pesticides and Pollution*, London: Collins.

Messenger, K. G. (1968) 'A railway flora of Rutland', *Proceedings of the Botanical Society of the British Isles*, vol. 7, pp. 325–44.

Miles, J. (1979) *Vegetation Dynamics*, London: Chapman and Hall.

Mitchell, F. (1963) 'Morainic ridges on the floor of the Irish Sea', *Irish Geography*, vol. 4, pp. 335–44.

—— (1965) 'Littleton Bog, Tipperary: an Irish vegetational record', *Geological Society of America, Special Paper No. 84*, pp. 1–16.

—— (1976) *The Irish Landscape*, London: Collins.

—— (1986) *The Shell Guide to Reading the Irish Landscape*, Dublin: Country House.

Montford, H. M. (1970) 'The terrestrial environment during the Upper Cretaceous and Tertiary times', *Proceedings of the Geologists' Association*, vol. 81, pp. 181–204.

Montier, D. J. (1977) *Atlas of Breeding Birds of the London Area*, London: Batsford.

Nature Conservancy Council (1977) *Nature Conservation and Agriculture*, London: Nature Conservancy Council.

—— (1984) *Nature Conservation in Great Britain*, Peterborough: Nature Conservancy Council.

—— (1986) 'Return of the large blue butterfly', *Topical Issues*, vol. 2, p. 1.

—— (1987) *Changes in the Cumbrian Countryside*, First report of the National Countryside Monitoring Scheme, Peterborough: Nature Conservancy Council.

—— (1988a) *The Flow Country: the Peatlands of Caithness and Sutherland*, Peterborough: Nature Conservancy Council.

—— (1988b) *Sites of Special Scientific Interest*, Peterborough: Nature Conservancy Council.

Nelson, E. C. (1978) 'Tropical drift fruits and seeds on coasts in the British Isles and western Europe, 1. Irish beaches', *Watsonia*, vol. 10, pp. 103–12.

Nelson, G. (1978) 'From Candolle to Croizat: comments on the history of biogeography', *Journal of the History of Biology*, vol. 11, pp. 269–305.

Newbigin, M. (1936) *Plant and Animal Geography*, London: Methuen.

Nilsson, T. (1983) *The Pleistocene*, Stuttgart: Ferdinand Enke.

Norton, T. A. (1978) 'Mapping species distributions as a tool in marine ecology', *Proceeding of the Royal Society of Edinburgh*, vol. 76B, pp. 201–13.

—— (ed.) (1985) *Provisional Atlas of the Marine Algae of Britain and Ireland*, Abbots Ripton: Institute of Terrestrial Ecology.

O'Sullivan, P. J. (1983) 'The distribution of the pine marten (*Martes martes*) in the Republic of Ireland', *Mammal Review*, vol. 13, pp. 39–44.

Owen, J. and Owen, D. F. (1975) 'Suburban gardens: England's Most Important Nature Reserve?', *Environmental Conservation*, vol. 2, pp. 53–9.

Packham, J. R. and Harding, D. J. L. (1982) *Ecology of Woodland Processes*, London: Edward Arnold.

Patterson, C. (1981) 'Biogeography: in search of principles', *Nature*, vol. 291, pp. 612–13.

—— (1983) 'Aims and methods in biogeography', in R. W. Sims, J. H. Price, and P. E. S. Whalley (eds) *Evolution, Time and Space: The Emergence of the Biosphere*, The Systematics Association special volume no. 23, London: Academic Press, pp. 1–28.

Pears, N. (1985) *Basic Biogeography*, London: Longman.

Pennington, W. (1969) *The History of the British Vegetation*, London: The English Universities Press.

Perring, F. H. (1963) 'Data-processing for the Atlas of the British Flora', *Taxon*, vol. 12, pp. 183–90.

—— (1968a) 'The Biological Records Centre and Irish plant and animal records', *Irish Naturalist Journal*, vol. 16, pp. 71–3.

—— (ed.) (1968b) *Critical Supplement to the Atlas of the British Flora*, London: Nelson.

—— (1969) 'The botanical importance of roadside verges', in J. M. Way (ed.) *Road Verges, Their Function and Management*, Monks Woods: Nature Conservancy.

—— (1970) 'The last seventy years', in F. H. Perring (ed.) *The Flora of a Changing Britain*, Faringdon: Classey, pp. 125–35.

—— and Walters, S. M. (1962) *Atlas of the British Flora*, London: Nelson.

Peterken, G. (1981) *Woodland Conservation and Management*, London: Chapman and Hall.

Pianka, E. R. (1983) *Evolutionary Ecology*. 3rd Edition, New York: Harper and Row.

Pigott, C. D. (1975) 'Experimental studies of the influence of climate on the geographical distribution of plants', *Weather*, vol. 30, pp. 82–90.

—— and Huntley, J. P. (1981) 'Factors controlling the distribution of *Tilia cordata* at the northern limits of its geographical range. III. Nature and causes of seed sterility', *New Phytologist*, vol. 87, pp. 817–39.

Pollard, E. M., Hooper. M. D. and Moore, N. W. (1974) *Hedges*, London: Collins.

Pomerol, C. (1982) *The Cenozoic Era*, Chichester: Ellis Horwood.

Post, L. von (1916) 'Forest tree pollen in south Swedish peat bog deposits', Lecture to the 16th Convention of Scandinavian naturalists in Kristiana (Oslo). Translated by M. B. Davis and K. Faegri, *Pollen Spores*, vol. 9., pp. 375–401.

Praeger, R. L. (1896) 'On the botanical subdivision of Ireland', *Irish Naturalist*, vol. 5, pp. 29–39.

—— (1901) *Irish Topographical Botany*, Published as *Proceedings of the Royal Irish Academy*, vol. 3.

—— (1906) 'A simple method of representing geographical distribution', *Irish Naturalist*, vol. 17, pp. 88–94.

—— (1932) 'Recent views bearing on the problem of the Irish fauna and flora', *Proceeding of the Royal Irish Academy*, vol. 41B, pp. 125–45.

—— (1934) *The Botanist in Ireland*, 1974 reprint, Wakefield: EP Publishing.

—— (1939) 'The relations of the flora and fauna of Ireland to those of other countries', *Proceedings of the Linnean Society*, vol. 151, pp. 192–213.

Preece, R. C. , Coxon, P., and Robinson, J. E. (1986) 'New biostratigraphic evidence of the Post-glacial colonization of Ireland and for Mesolithic forest disturbance', *Journal of Biogeography*, vol. 13, pp. 487–509.

Preston, F. W. (1962) 'The canonical distribution of commonness and rarity: Part I', *Ecology*, vol. 43, pp. 185–215. Part II, vol. 43, pp. 410–32.

Rackham, O. (1976) *Trees and Woodland in the British Landscape*, London: Dent.

—— (1980) *Ancient Woodland: its History, Vegetation and Uses in England*, London: Edward Arnold.

Ratcliffe, D. A. (1968) 'An ecological account of Atlantic bryophytes in the British Isles', *New Phytologist*, vol. 67, pp. 365–439.

—— (1974) 'Ecological effects of mineral exploitation in the United Kingdom and their significance to nature conservation', *Proceedings of the Royal Society of London: A*, vol. 339, pp. 355–72.

—— (ed.) (1977) *A Nature Conservation Review*, Cambridge: Cambridge University Press.

—— (1984) 'Post-medieval and recent changes in British vegetation: the culmination of human influence', *New Phytologist*, vol. 98, pp. 73–100.

Raunkiaer, C. (1934) *The Life Forms of Plants and Statistical Plant Geography*, Oxford: Clarendon Press.

Reid, E. M. (1949) 'The Late-Glacial flora of the Lea Valley', *New Phytologist*, vol. 28, pp. 245–52.

—— and Chandler, M. E. J. (1933) *The London Clay Flora*, London: British Museum.

Ridley, H. N. (1930) *The Distribution of Plants Throughout the World*, Ashford: Reeve.

Robinson, M. (1986) 'The extent of farm underdrainage in England and Wales, prior to 1939', *Agricultural History Review*, vol. 34, pp. 79–85.

Roden, C. (1979) 'The vascular flora and vegetation of some islands of Loch Corrib', *Proceedings of the Royal Irish Academy*, vol. 79, pp. 223–34.

Royal Society for the Protection of Birds (1984) *Hill Farming and Birds: a Survival Plan*, Sandy: Royal Society for the Protection of Birds.

Ruddiman, W. F. and McIntyre, A. (1981) 'The North Atlantic during the last deglaciation', *Palaeogeography, Palaeoclimatology, Palaeoecology*, vol. 35, pp. 145–214.

Salisbury, E. J. (1932) 'The East Anglian Flora', *Transactions of the Norfolk and Norwich Natural History Society*, vol. 13, pp. 191–263.

—— (1939) 'Ecological Aspects of Meteorology', *Quarterly Journal of the Royal Meteorological Society*, vol. 65, pp. 337–58.

Sargent, C. (1984) *Britain's Railway Vegetation*, Abbots Ripton: Institute of Terrestrial Ecology.

Scott, N. E. and Davison, A. W. (1982) 'De-icing salt and the invasion of road verges by maritime plants', *Watsonia*, vol. 14, pp. 41–52.

Seaward, D. R. (1982) *Sea Area Atlas of the Marine Molluscs of Britain and Ireland*, Shrewsbury: Nature Conservancy Council.

Seaward, M. R. D. and Hitch, C. J. B. (1982) *Atlas of the Lichens of the British Isles*, Cambridge: Institute of Terrestrial Ecology.

Seddon, B. (1971) *Introduction to Biogeography*, London: Duckworth.

Sharrock, J. T. R. (1976) *The Atlas of Breeding Birds of Great Britain and Ireland*, Berkhamstead: Poyser.

Sheail, J. (1980) *Historical Ecology: The Documentary Evidence*, Abbots Ripton: Institute of Terrestrial Ecology.

Shimwell, D. (1971) *Description and Classification of Vegetation*, London: Sidgwick & Jackson.

Simberloff, D. S. (1976) 'Experimental zoogeography of islands: effects of island size', *Ecology*, vol. 57, pp. 629–48.

Simkins, J. and Williams, J. I. (1981) *Advanced Biology*, London: Bell & Hyman.

Simmons, I. G. (1974) *The Ecology of Natural Resources*, London: Edward Arnold.

—— (1980) 'Biogeography', in E. H. Brown (ed.) *Geography Yesterday and Tomorrow*, Oxford: Oxford University Press, pp. 146–66.

—— and Tooley, M. J. (eds) (1981) *The Environment in British Prehistory*, London: Duckworth.

Simpson, G. G. (1965) *The Geography of Evolution*, Philadelphia: Chilton.

Southward, A. J. (1958) 'Note on the temperature tolerances of some intertidal animals in relation to environmental temperatures and geographical distribution', *Journal of the Marine Biological Association*, vol. 37, pp. 49–66.

—— and Southward, E. C. (1978) 'Recolonisation of rocky shores in Cornwall after use of toxic dispersants to clean up the Torrey Canyon spill', *Journal of the Fisheries Research Board of Canada*, vol. 35, pp. 682–706.

Southwood, T. R. E. (1977) 'Habitat, the template for ecological strategies', *Journal of Animal Ecology*, vol. 46, pp. 337–65.

Stamp, L. D. (1962) 'A geographer's postscript', in D. Nichols (ed.) *Taxonomy and Geography*, The Systematics Association special volume no. 4. London: Academic Press, pp. 153–8.
—— (1969) *Nature Conservation in Britain*, London: Collins.
Stoddart, D. R. (1965) 'Geography and the ecological approach. The ecosystem as a geographical principle and method', *Geography*, vol. 50, pp. 241–51.
—— (1977) 'Biogeography', *Progress in Physical Geography*, vol. 1, pp. 537–43.
—— (1981) 'Biogeography: dispersal and drift', *Progress in Physical Geography*, vol. 5, pp. 575–90.
—— (1986) *On Geography*, Oxford: Blackwell.
Stowe, T. J. (1987) 'The management of sessile oakwoods for pied flycatchers', *RSPB Conservation Review*, no. 1, pp. 78–83.
Stuart, A. J. (1982) *Pleistocene Vertebrates in the British Isles*, London: Longman.
Stuttard, P. and Williamson, K. (1971) 'Habitat requirements of the nightingale', *Bird Study*, vol. 18, pp. 9–14.
Sundene, O. (1962) 'The implications of transplant and culture experiments on the growth and distribution of *Alaria esculenta*', *Nytt Magasin for Botanikk*, vol. 9, pp. 155–74.
Sutton, S. L. (1972) *Woodlice*, London: Ginn.
Tait, R. V. (1981) *Elements of Marine Ecology*, London: Butterworths.
Tansley, A. G. (1935) 'The use and abuse of vegetational concepts and terms', *Ecology*, vol. 16, pp. 284–307.
—— (1939) *The British Islands and their Vegetation*, London: Cambridge University Press.
Taylor, J. A. (1976) 'Upland climates', in T. J. Chandler and S. Gregory, *The Climate of the British Isles*, London: Longman, pp. 264–88.
—— (ed.) (1984) *Themes in Biogeography*, London: Croom Helm.
Taylor, J. W. (1907) *Monograph of the Land and Freshwater Mollusca of the British Isles*, Leeds: Taylor Bros.
Taylor, L. R. (1986) 'Synoptic dynamics, migration and the Rothamsted Insect Survey', *Journal of Applied Ecology*, vol. 55, pp. 1–38.
——, French, R. A. and Woiwod, I. P. (1978) 'The Rothamsted Insect Survey and the urbanisation of land in Great Britain', in G. W. Frankie and C. S. Koehler, *Perspectives in Urban Entomology*, New York: Academic Press, pp. 31–65.
Teagle, W. G. (1978) *The Endless Village*, Shrewsbury: Nature Conservancy Council.
Tivy, J. and O'Hare, G. (1980) *Man and the Ecosystem*, Edinburgh: Oliver & Boyd.
Topham, P. B., Alphey, T. J. W., and Shaw, R. (1983) 'Distribution mapping at the national level for the European plant-parasitic nematode survey', *Nematologica Mediterranea*, vol. 11, pp. 1–11.
Tooley, M. (1981) 'Methods of reconstruction', in I. G. Simmons and M. J. Tooley (eds) *The Environment in British Prehistory*, London: Duckworth, pp. 1–49.
Tout, D. (1976) 'Precipitation', in T. J. Chandler and S. Gregory, *The Climate of the British Isles*, London: Longman, pp. 96–129.
Troels-Smith, J. (1960) 'Ivy, mistletoe and elm climatic indicators – fodder plants', *Danmarks Geologiske Undersogelse, IV*, vol. 4, pp. 1–32.
Tuhkanen, S. (1980) 'Climatic parameters and indices in plant geography', *Acta Phytogeographica Suecica*, vol. 67, pp. 9–109.
Turner, J. (1970) 'Post-Neolithic disturbance of British vegetation', in D. Walker and R. G. West (eds) *Studies in the Vegetational History of the British Isles*, Cambridge: Cambridge University Press, pp. 97–116.

Turner, R. G. (1986) 'Why are there so few butterflies in Liverpool? Homage to Alfred Russel Wallace', *Antenna*, vol. 10, pp. 18–24.
Turrill, W. B. (1959) 'Plant geography' in W. B. Turrill (ed.) *Vistas in Botany*, London: Pergamon, pp. 172–229.
Tyler, S. (1987) 'River birds and acid water', *RSPB Conservation Review*, vol. 1, pp. 68–71.
Udvardy, M. D. F. (1969) *Dynamic Zoogeography*, New York: Van Nostrand Reinhold.
Usher, M. B. (1986a) 'Invasibility and wildlife conservation: invasive species on nature reserves', *Proceedings of the Royal Society of London: B*, vol. 314, pp. 695–710.
—— (ed.) (1986b) *Wildlife Conservation Evaluation*, London: Chapman and Hall.
Varley, G. C. (1970) 'The concept of energy flow applied to a woodland community', in A. Watson (ed.) *Animal Populations in Relation to Their Food Resources*, Oxford: Blackwell, pp. 389–405.
Vink, A. P. A. (1983) *Landscape Ecology and Land Use*, London: Longman.
Walker, D. (1970) 'Direction and rate in some British post-glacial hydroseres', in D. Walker and R. G. West (eds) *Studies in the Vegetational History of the British Isles*, Cambridge: Cambridge University Press, pp. 117–41.
Wallace, A. R. (1876) *The Geographical Distribution of Animals, with a Study of the Living and Extinct Faunas as Elucidating the Past Changes of the Earth's Surface*, London: Macmillan.
—— (1880) *Island Life: or, the Phenomena and Causes of Insular Faunas and Floras, Including a Revision and Attempted Solution of the Problem of Geological Climates*, London: Macmillan.
Walters, S. M. (1957) 'Distribution maps of plants – an historical study', in J. E. Lousley (ed.) *Progress in the Study of the British Flora*, Botanical Society of the British Isles Conference Report, pp. 89–96.
—— (1978) 'British endemics', in H. E. Street (ed.) *Essays in Plant Taxonomy*, London: Academic Press, pp. 263–74.
Watson, H. C. (1832) *Outlines of the Geographical Distribution of British Plants; Belonging to the Division of Vasculares or Cotyledones*, Edinburgh.
—— (1835) *Remarks on the Geographical Distribution of British Plants*, London: Longman.
—— (1847–59) *Cybele Britannica: or, British Plants and their Geographical Relationships*, London: Longman.
—— (1873–4) *Topographical Botany*, London: Ditton.
Watts, W. A. (1985) 'Quaternary vegetation cycles', in K. J. Edwards and W. P. Warren (eds) *The Quaternary History of Ireland*, London: Academic Press, pp. 155–83.
Webb, D. A. (1980) 'The biological vice–counties of Ireland', *Proceeding of the Royal Irish Academy*, vol. 80B, pp. 179–96.
—— (1983) 'The flora of Ireland in its European context', *Journal of Life Sciences, Royal Dublin Society*, vol. 4, pp. 143–60.
West, R. G. (1970) 'Pollen zones in the Pleistocene of Great Britain and their correlation', *New Phytologist*, vol. 69, pp. 1179–83.
—— (1980a) 'Pleistocene forest history in East Anglia', *New Phytologist*, vol. 85, pp. 571–622.
—— (1980b) *The Pre-Glacial Pleistocene of the Norfolk and Suffolk Coasts*, Cambridge: Cambridge University Press.
——, Dickson, C. A., Catt, J. A., Weir, A. H. and Sparks, B. W. (1974) 'Late Pleistocene deposits at Wretton, Norfolk. II. Devensian deposits', *Philosophical Transactions of the Royal Society of London, B*, vol. 267, pp. 337–420.

Whittaker, R. H. (1953) 'A consideration of climax theory: climax as population and pattern', *Ecological Monographs*, vol. 23, pp. 41–78.

Williams, C. B. (1943) 'Area and the number of species', *Nature*, vol. 152, pp. 264–7.

—— (1953) 'The relative abundance of different species in a wild population', *Journal of Animal Ecology*, vol. 22, pp. 14–31.

Williamson, M. (1981) *Island Populations*, Oxford: Oxford University Press.

Willis, J. C. (1922) *Age and Area: a Study in Geographical Distributions and Origin of Species*, Cambridge: Cambridge University Press.

Wilmott, A. J. (1930) 'Concerning the history of the British flora', *Société de Biogéographie*, vol. 3, pp. 163 *et seq.*

—— (1935) 'Evidence in favour of survival of the British flora in glacial times', *Proceedings of the Royal Society, B*, vol. 118, pp. 197–241.

Wilson, A. (1956) *The Altitudinal Range of British Plants*, Arbroath: Buncle.

Woodward, F. I. (1976) 'The climatic control of the altitudinal distribution of *Sedum rosea* (L.) Scop. and *S. telephium* L. 2. The analysis of plant growth in controlled environments', *New Phytologist*, vol. 74, pp. 335–48.

—— and Pigott, C. D. (1976) 'The climatic control of the altitudinal distribution of *Sedum rosea* (L.) Scop. and *S. telephium* L. 1. Field observations', *New Phytologist*, vol. 74, pp. 323–34.

Woodwell, G. M. (1970) 'The energy cycle of the biosphere', *Scientific American*, vol. 223, pp. 64–74

Index

Common and specific names are listed here for organisms discussed in the text. Page entries are under the Latin names. If no common name is available an indication of the organisms identity is given in brackets after the Latin name. An asterisk refers to a test figure.

Skokholm (Pembrokeshire) 75, 125
Skomer (Pembrokeshire) 238
skylark, *see Alauda arvensis*
slender naiad, *see najas flexilis*
slender rush, *see Juncus filiformis*
Slieve League (Co. Donegal) 202
small mountain ringlet (a butterfly), *see Erebia epiphron*
small teasel, *see Dipsacus pilosus*
small-leaved lime, *see Tilia cordata*
small white (a butterfly), *see Pieris rapae*
snakelocks anemone, *see Anemone sulcata*
Snowdon eyebright, *see Euphrasia rivularis*
snowfall and liverwort distribution 152
Society for the Promotion of Nature Reserves 237
soils: acidity or pH 46; calcareous rendzinas 42*, 46; calcareous soils 42*; climatic relationships 46–7; edaphic factors 40; environment 40; geographical distribution 41*; gleying 47; leaching 44; nutrient availability and pH 45*; podsolic soils 42*, 46; profiles 42*; structure 44*; texture 43*
soil-vegetation complex and biogeography 13
solar radiation: levels 90; usage by leaves 91*
Solaster endica 174
Solbrig, O. T. 81
Solea solea 135
Solomon's seal, *see Polygonatum multiflorum*
Somerset Moors and Levels: drainage 233, 234*
Sonchus asper 217
S. oleraceus 217
southern floristic element 159
southern marsh orchid, *see Dactylorhiza praetermissa*
Southward, A. J. 84–5, 141, 252
Southward, E. C. 252
Southwood, T. R. E. 79
sow-thistles, *see Sochus* spp.
Spanish catchfly, *see Silene otites*
sparrowhawk, *see Accipiter nisus*
Spartina alternifolia 75
  S. anglica 75
  S. maritima 75

S. townsendii 75
spatial correlation, problems of interpretation 129
speciation 75
species: definition 74; diversity index 267; equilibrium levels 125; size in urban areas 268
Species–abundance distributions 122
species–area equation 120
Sphagnum bog and hydrosere successions 99*
spiked rampion, *see Phyteuma spicatum*
spiny lobster, *see Palinurus vulgaris*
spiny starfish, *see Marthasterias glacialis*
Spiranthes romanzoffiana 162
spotted hyaena, *see Crocuta crocuta*
spring whitlow grass, *see Erophila verna*
spruce, *see Picea abies*
squirrel spp. and competitive exclusion 119
Stachys germanica 232
Stamp, L. D. 60, 225
standing stock, definition 92
Stanlow Marsh (Merseyside) 256
Statutory Nature Reserves (Eire) 239
stemless thistle, *See Cirsium acaulon*
Steno's horse, *see Equus stenonis*
sticky groundsel, *see Senecio viscosus*
stiff sedge, *see Carex bigelowii*
Stoddart, D. R. 2, 10, 13, 18
Stongylocentrotus drobachiensis (a sea-urchin) 174
Stowe T. J. 88
Strachan, I. 197
straight-tusked elephant, *see Palaeoloxodon antiquus*
Straits of Dover: Lower Quaternary land-bridge 183; Middle Pleistocene 190; post-glacial flooding 211
strawberry tree, *see Arbutus unedo*
Streptopelia decaocto 215
striated catchfly, *see Silene conica*
Stuart, A. J. 182, 188–9, 197
stubble burning and partridge decline 232
Studdard, P. 111
subclimax 98
sublittoral epibenthic habitats, classification, 78
succession 9; autogenic 98; Clements' six phases 98; climax 97; primary 98; secondary 98